Beyond the Energy Crisis

Beyond the Energy Crisis

JOHN MADDOX

McGraw-Hill Book Company
New York St Louis San Francisco

Library of Congress Cataloging in Publication Data

Maddox, John Royden, 1925
Beyond the energy crisis

1. Energy policy
2. Petroleum industry and trade. I. Title

HD9502.A2M33 333.7 75–8922

ISBN 0–07–039430–x

Printed in Great Britain by
The Anchor Press Ltd
Tiptree, Essex, England

Contents

Author's notes

Throughout this book, quantities are measured (wherever appropriate) in metric tons, sometimes written *tonnes*. The following energy equivalents have also been used: 1 ton of oil = 1.6 tons of coal = 7.1 barrels of oil (42 US gallons or 33.5 Imperial gallons). The complete fission of 1 ton of uranium yields the energy equivalent of 3 million tons of crude oil.

I am grateful to politicians, government officials and managers in the energy industries in many countries for stimulating conversations. Books and documents which I have found especially helpful are referred to in the bibliographical notes. I am especially grateful to the publishers for their forbearance, to my wife Brenda Maddox for her encouragement and to my secretary, Mary Sheehan, for her patience.

Introduction

In the industrial history of the West, the 1960s were probably the most prosperous decade on record. The chances are that the 1970s will look very different in retrospect, and at least a part of the reason will be the sharp increase in the price of oil sold by the oil-producing states. This transition is already known as the 'energy crisis', and the oil-producing states are blamed for having visited a whole range of troubles on the world. In reality, however, what is happening today has much wider ramifications, and, indeed, if the word 'crisis' is applicable at all, it should be applied not narrowly to energy but more widely to an inevitable change in the political and economic relationship between the oil-producing and the oil-consuming states.

The bare facts are as follows. In the past few years, but particularly during the 1960s, petroleum from the Middle East, and to a lesser extent from such other oil-producing states as Venezuela, Libya and Indonesia, was plentiful and cheap. During that decade, Western Europe and Japan came to rely for their supplies of energy on imports of petroleum. In Europe, oil consumption increased from 612 million tons in 1960 to 1041 million tons in 1970. In Japan, oil consumption trebled during the decade. By the end of the 1960s, the United States, previously self-sufficient in petroleum, had become a substantial importer for the first time. Throughout the 1960s, the price of petroleum on the international markets had been stable and even falling, with the oil-producing states receiving royalties on their oil exports from the oil companies concerned.

Back in 1960, however, the chief oil-exporting states, most of them in the Persian Gulf and North Africa, had formed the Organization of Petroleum Exporting Countries (OPEC), chiefly in response to the decision of some of the American oil companies operating in the Persian Gulf to reduce the price of oil on which the producer governments' revenues were calculated. The decision was outwardly sensible enough – the potential supply of oil from the Middle East was much greater than the then demand – but since one consequence was to reduce the revenues of the oil-exporting countries, it was a tactless move to say the least.

During the 1950s the governments who were to form OPEC had managed to shake off the traditional system of rewards for oil produced, which had usually been based on a fixed payment of roughly $0.15 a barrel. One government after another had managed to establish the principle that its revenue should be based on a notional profit sharing – and the oil concessions and refineries in Iran (then Persia) were nationalized in 1953 when British Petroleum declined to extend this principle to its Persian landlords. The fact that oil production in Persia was reduced to virtually nothing for two years without causing any noticeable slackening in the pace of growth of oil consumption in Western Europe is a measure of how, then, the demand for oil failed to match the potential supply. The profit-sharing formula of the 1950s depended on the creation of the posted prices for oil from the Middle East. The principle was for the companies to pay a royalty (usually 12½ per cent of the posted price) and also a proportion (originally a half but later 55 per cent) of the difference between the posted price and production cost. Much of the haggling between OPEC and the oil companies in the 1960s concerned issues such as the treatment of the royalty payment as a part of production cost, and the allowance to be made for the cost to the companies for marketing and distribution.

To begin with, OPEC was weak and tentative. Its commercial power rested, and still rests, not so much on the size of its oil reserves as on oil's cheapness relative to other sources of energy, and OPEC was slow to recognize its own strength. From the start, however, it saw that a producers' cartel needed to be able to regulate the volume of production if it was to be effective, but its members could not agree among themselves as to how this should be done. The states with the largest reserves argued that production should be proportional to proved reserves. The most populous states, Iran for example, argued that production should be proportional to population. Others sought to preserve the existing relationship among the OPEC members, asking that production should always move up or down by the same proportion in each oil-producing state.

The early 1960s were also bad times for setting up an oil-producers' cartel. Oil was plentiful, the great international fleets of tankers were being founded, and the oil companies were beginning to enjoy the economies of scale that came from the building of large modern refineries. During the 1960s, under increasingly strong pressure from OPEC, the receipts of the oil-producing states rose steadily, from $0.40 a barrel in the early 1960s to twice that, and in some cases to just under $1.00 a barrel, by 1970. It seems ironical today to recollect how, during this period, the tensions between governments and their concessionaires often took the form of attempts by the governments to persuade the companies that production – and thus revenues – should be increased.

By the end of the decade, however, OPEC was a tougher animal. Its membership was extended with the accession of Indonesia, Venezuela and Nigeria. The producing states were increasingly resentful of several aspects of the international petroleum trade – the high profits of the oil companies, the large revenues collected by consuming governments in excise duties and the relative cheapness of Middle East oil. Throughout the 1960s, OPEC had become steadily more sure of itself; its common interest in its profitability was strong enough to override the disparities between the members – left-wing Iraq and vigorously anti-communist Saudi Arabia have, after all, a common interest in maximizing oil revenues, as do the Arab members of OPEC and the ostentatiously non-Arab Iran.

By 1968 OPEC's claim on the oil companies had been made plain. In June, an OPEC resolution spelled out the objectives of the OPEC governments. As far as possible, exploration and development were to be carried out by the governments themselves. Where oil companies were awarded concessions, goverments should be free to acquire a 'reasonable' measure of participation 'on the grounds of the principle of changing circumstances' – one of the principles that would be used between 1971 and 1973 to secure the renegotiation of the basic agreement between companies and governments. Schemes under which concessionary companies were required to hand back acreage to the governments were to be accelerated. Governments were to fix posted prices and were to be free to change these in the light of changing prices for manufactured goods. Tax fixing was to be the prerogative of producing governments, and if negotiations with the companies should be unsuccessful, the governments were to be free to fix taxes so that the earnings of the oil companies should be both reasonable and sufficient to make it worthwhile for the companies to shoulder the risks of exploration and development.

From this point on, it has at least been clear that the heyday of the concessionary companies was over. In the annual report of British Petroleum (BP) for 1970, Sir Eric Drake, the chairman, said that the oil companies had become 'tax gatherers' for the OPEC governments. The next concerted demand was to be for participation, and it culminated in the agreement between OPEC governments and the oil companies signed in Tehran in February 1971. This agreement fixed new prices and price differentials, with a built-in escalation factor to take account of inflation in the prices of manufactured goods imported from the West. But its chief importance was a formula for the acquisition by the producing governments of a controlling interest in the oil companies to which concessions had been let, usually in the decade to 1981.

The Tehran agreement, hailed at the time by many in the West as an enlightened document, has, for practical purposes, been honoured in the breach, not the observance. Apart from minor adjustments to account for the devaluation of the dollar (Geneva, 1972), the agreement was followed by a sequence of demands for immediate control.

Iran did most to breach the dyke. After negotiations between the government and the Iranian Oil Consortium (principally BP and Shell) had dragged on through 1972, the Shah faced the companies with the prospect of legislation. By July 1973, the consortium became, by an Act of the Iranian parliament, two separate organizations: (a) a service company employed by the National Iranian Oil Company to look for new oil, develop existing oil fields, help operate existing refineries and petrochemicals plants and train Iranian staff: and (b) a commercial organization (the consortium) to buy back from the National Iranian Oil Company specified quotas of crude oil and petroleum products. In principle, an increasing share of Iranian oil is at the disposal of the National Iranian Oil Company. In the early months of 1973, the members of the Iran Consortium were triumphant over the prospect of handling not merely the share of Iranian crude to which they are entitled but the National Iranian Oil Company's share as well. But in fact, thanks partly to the scramble for oil that developed in the closing months of 1973, the Iranians have shown great skill in marketing their own products. By the end of 1973, the National Iranian Oil Company was making no secret of its belief that it should immediately have a larger share of its production to market as it chose. During 1974, oil producers such as Kuwait and Saudi Arabia followed Iran in this demand for full control, and the principle has been established that the international oil companies will, in future, be at most the paid agents of the oil-producing states.

Why has this course of events come as such a shock to the oil consumers of the West? The truth is that nobody need have been surprised. From the mid-1960s the oil companies were uttering stern warnings about their increasingly uneasy relationship with OPEC countries. Some governments did see the writing on the wall. In Japan, for example, the signs of impending change stimulated the movement by Japanese oil companies and financial groups to finance a variety of projects in the Middle East, including oil refineries and gas liquefaction plants as well as copper mines. The calculation, falsified by events, was that the provision of the capital needed to increase industrialization in countries such as Kuwait would secure a reasonable share of oil production. And there were signs even in the United States that the Administration appreciated the likely course of changing circumstances. In April 1971, for example, President Richard M. Nixon, in the first of a series of messages to Congress on the theme of energy, gave an accurate forecast of what lay ahead and announced a policy of self-sufficiency in energy. Two further years went by, however, before the United States began to implement its declared policy on energy. Only the embargo of most Arab members of OPEC after the Yom Kippur war persuaded the administration that something really must be done. We still wait to hear what that something will be.

In several ways, OPEC has thus been resoundingly successful. It has won more income for its members, and it has been able to establish the overriding political point that the oil consumers can ignore those developing countries which are sources of important raw materials only at their peril. No wonder that the producers of copper, bauxite and other raw materials have welcomed OPEC's example, though it remains uncertain whether they will be able to profit as much from it as they hope.

But OPEC has also overreached itself. The going price of oil through much of 1974 was roughly $10.00 a barrel in the Persian Gulf, and rather more when delivered to the markets of the West. This price is substantially above the probable price of other sources of energy – not merely as compared with the oil shales of North America or nuclear electricity based on uranium as a fuel, but also with the reserves of conventional petroleum now being exploited in many other places throughout the world.

What are the consequences for the oil consumers? The most immediate is that the oil-consuming governments have lost the sense – illusory since 1960 – that the oil reserves of the OPEC countries could be counted on as a safe extension of their own supplies of energy. Indeed, the selective oil embargo imposed by the Arab members of OPEC between October 1973 and January 1974 was a proof that, in present circumstances, while the oil market remains a sellers' market, there is every possibility that supplies of oil could in future be withheld for political reasons. This prospect is not merely uncomfortable but intolerable. The urgent question facing Western governments is how this prospect can be diminished without sacrificing too much of the principle that the oil consumers should be free to pursue those international policies which they consider to be just, and in particular that the United States should remain free to insist on the right of Israel to exist. Old-fashioned diplomacy has no doubt much to contribute, but, in the long run, the best assurance against political embargoes on oil is for the commercial power of OPEC to be weakened, which implies that the industrialized countries of the West must urgently develop the alternative sources of energy that are available.

The new high price of oil decreed by OPEC since the beginning of 1974 has had other painful consequences, though here it is important not to exaggerate the scale of what has happened. The total annual revenues of the OPEC states from the export of oil during the later 1970s are likely to exceed $100 000 million, much of which will remain unspent. Arithmetically, these sums are very large, but the oil revenues of the OPEC states during 1974 were between 1 and 2 per cent of the Gross National Product of their customers for oil. On the face of things, even if the OPEC states were able in the immediate future fully to consume the goods and services which their oil revenues might buy, the oil consumers might expect to pay their bills merely by forgoing rather less than a year's normal increase of their own consumption.

Unfortunately, the increase of the revenues of oil-producing states has other and less obvious consequences. First, the increased price of oil has come at a time when inflation is firmly established in many of the industrialized countries of the West. Secondly, the financial institutions of the West, in which the unspent part of the OPEC revenue must perforce be invested, are not well suited to the handling of such large amounts of credit: (a) inflation has been accelerated; and (b) given the tendency for the surplus funds to find their way into the hands of the safest bankers, the external financial problems of the weakest oil consumers (conspicuously Italy and Britain) have been sharpened. Thirdly, because the new high price of oil forces on the oil consumers the need to invest in alternative sources of energy (or, which comes to the same thing, to invest in means of reducing consumption), they find themselves faced with the need to increase their own capital investment at a time when high interest rates persuade them in the opposite direction. But the increased price of oil has also coincided with a general tendency for the price of raw materials of all kinds to increase, which is, in itself, partly a consequence of the growing awareness of developing countries with exportable raw materials that they, like the members of OPEC, are entitled to a larger share of the benefit from their activities and resources.

In this sense, what has happened between the oil consumers and OPEC is a part of the general realignment of the relationship between the industrialized countries of the West (and the East, for that matter). During 1974, the signs of this essentially political process, unpredictable in its outcome but still far from complete, included the Special Assembly of the United Nations on a 'new world economic order' and the explosion of a nuclear device by India in the early summer. The developing countries, often more seriously affected by the increased price of oil than the industrialized countries, have actually been careful not to protest too loudly at how OPEC's higher prices have influenced their affairs. For them, OPEC is not an affront but an example.

The book which follows is largely an attempt to put these events into perspective. It is important to realize that the changed pattern of Western industry that will undoubtedly follow from the increased price of oil will be, in character, no different from the kinds of changes that have come about in the historical past. The difficulty is merely that the changes now in prospect are enforced by external circumstances, not a part of the natural evolution of industrial society. In the long run, however, there is no reason why the increased price of oil should prove an intolerable burden for the West. Everything will depend on the skill with which it can somehow create the circumstances in which technical innovation can be welcomed.

In the short run, over the next few years, there are constructive steps to be taken. While there is no prospect that the price of OPEC oil will ever again be reduced to what it used to be in the 1960s, there is good reason to suppose that the prices now effective are higher than they would be if OPEC had not acquired a monopoly of cheap oil. The implied calculation is that alternative sources of energy – oil from other oil fields still undeveloped, coal and even nuclear power – will become relatively cheaper and more exploitable. In round numbers, there are strong grounds for thinking that if the oil consumers were free to choose between OPEC oil and these alternatives, the price of oil would settle down at roughly $7.00 a barrel. Ultimately, this is what it will amount to. But the process could be hastened, and the uncomfortable features of the OPEC monopoly softened, given a sharp if temporary reduction of oil consumption by the oil consumers. Reducing consumption by about 15 per cent would reduce the demand for OPEC oil by roughly a quarter and would, in the process, sharpen the present conflict of interest between those oil consumers who need their revenues from oil and those who would still find it necessary to deposit much of it with Western banks.

There are also the strongest possible reasons why the oil consumers should re-examine their policies on the production and use of energy, and set out deliberately to make these conform more accurately with the elementary principles of economics. As will be seen (Chapter 6), much of the responsibility for the upheaval of the past few years can be blamed on the ways in which the oil consumers have followed domestic policies that distort the natural pattern of consumption. In the process, they have squandered economic resources, not to mention energy. Some of the ways in which energy from various sources was marketed in the United States and Europe at prices widely different from those that would have obtained in a free market for energy survived even the sharp increase of the price of oil in 1974.

To say this is merely to imply that the old-fashioned principles of economics have a crucial part to play in making sure that resources are used efficiently. Price should be the moderator of the relationship between supply and demand, and devices which depart from this simple principle are, of necessity, devices for ensuring that resources will be wasted. But this simple principle does not of itself imply that governments have no part to play, or that the production and pricing of different kinds of energy should be exclusively the responsibility of commercial companies. It is merely that if governments choose to override the profit motive, they should do so rationally. Their record over the past few decades of this century has not been conspicuously successful.

I
Change upon change

The rapid increase in the price of oil between 1970 and 1974 will ultimately have a lasting effect on the pattern of energy consumption in the West. That much is beyond dispute. Put simply, if the increase of oil prices remains unchallenged, oil consumers will be persuaded to obtain energy from other sources. If energy as a whole becomes more expensive than other kinds of raw materials, they will be persuaded to make more economical use of energy. We can now see more clearly that the increase of oil prices should have been anticipated, that some of its consequences would, in any case, have been forced on us in the not too distant future by the depletion of conventional oil reserves, that alternative sources of energy at reasonable costs are indeed available and that the reaction of the industrialized world to recent events has been too passive for its own good and even for the long-term interests of the oil-producing states of OPEC.

A sober assessment of these issues must, however, depend on an appreciation of the nature of industrial history, which is not a process of steady evolution but, rather, an evolutionary process repeatedly punctuated by changes that are technically radical and socially painful. The consequences of the upheaval in the energy market now under way, however serious these may be, will in retrospect seem mild compared with some of the transformations of the pattern of industry over the past few centuries. With good management, they may yet be made to appear as little more than catalysts of beneficent changes which are, in any case, urgently needed.

The development of the new techniques by which the process of industrial change has been sustained has, in the past, had a profound influence on the character of society. The techniques of growing crops and of keeping domesticated animals made Neolithic society possible. The water wheel put its stamp on the social and economic structure of medieval Europe. The steam engine made possible the Industrial Revolution, and also determined the character of the industrial cities of Victorian England. What Daniel Bell calls the Post-Industrial Society, the hallmark of the prosperous 1960s, was made possible not merely by the rapid increase of productivity in manufacturing industry, but by the harvest of new techniques that flowed from the electronics laboratories of the industrialized world. That many of these important transformations have been brought about by new techniques for transforming energy from one form into another is no accident, but merely a proof that energy and the techniques for using it are essential ingredients of all societies.

Even before the increase of oil prices in 1974, the future supply of energy for the industrialized world was a conspicuous source of public anxiety in the West. The question was whether energy consumption could continue to increase, or whether the intensity of energy consumption in the impoverished countries of the world could ever approach that in the industrialized West without jeopardizing the natural resources on which industrial activity must depend. Such fears, based as they usually are on simple arithmetical projections into the future, are misplaced. They spring from the assumption that growth will be accomplished without qualitative change – that if the world's use of energy multiplies, say, ten times, the rate at which petroleum is used will also increase tenfold.

Such a course of events is exceedingly improbable. The industrial history of the past 10 000 years is an exciting tale of how new sources of energy have been used to supplement and sometimes replace the more traditional. And there is no reason to fear that the potential for innovation is now exhausted. On the contrary (see Chapter 3), the potential for change is now greater than it has ever been. The succession of innovations of technique for transforming energy from one form to another is thus a necessary starting point for the discussion of contemporary problems.

The first tools were mechanical devices, axes and spearheads, for making the energy of human muscles more effective. It is customary to suppose that these earliest stages of society made no deliberate use of energy, but that is to ignore the quite substantial amounts of energy consumed in the human diet, even in Paleolithic times. The average diet, at least for those communities that managed to survive, could not have been energetically much less generous than it is at present, roughly the equivalent of the energy in two-thirds of a barrel of petroleum.

Neolithic agriculture was a means of improving the quality of food and making the supply of it less chancy by the deliberate application of the energy of human muscles to the clearing of forests and the cultivation of the primitive fields created in this way. Cultivation was harder work than hunting and gathering, but the economic benefits were important if marginal to start with. It has been estimated that the average annual cost of cultivating a hectare (nearly 2.5 acres) of land by primitive methods is roughly 3000 man hours – one man's work for ten hours a day, six days a week, from one end of the year to the other. The minimum food requirements of a population working primitive agriculture has been estimated to be between 200 and 250 kilograms of grain a year for each person, children as well as adults. But the yield of grain from land cultivated by neolithic methods is unlikely to have exceeded one ton (1000 kilograms) a hectare except in exceptional circumstances. Thus the hardest working farmer could not have expected to support himself and more than three dependants exclusively by the techniques of primitive agriculture. He could hope to increase the economic value of his own hard work by exploiting the energy of draught animals only in circumstances so favourable that at least some grain could be spared for the animals, perhaps only when the amount of grain available reached twice the amount needed just to allow the human beings to survive. Only the careful selection of seeds, and the careful preservation of productive strains of grains from one generation to another, could have provided a foundation for the use of draught animals for clearing ground or pulling ploughs. The results, however, would have been a further increase of the productivity of the land, and a further

improvement of the human diet, both in amount and variety. Neolithic agriculture thus stands in relation to what preceded it as does the intensive agriculture of the twentieth century to the agriculture of Europe in the early eighteenth century, for modern agriculture is a means by which the yield of the land has been increased by supplementing the energy of farm labourers and their draught animals by the chemical energy of artificial fertilizers.

Primitive metallurgy, and especially the smelting of copper and tin (the constituents of bronze), required the use of one form of energy – that of charcoal – to extract metals from their ores, and must be likened to the creation of the great petrochemicals industry over the past thirty years, even though, when tools of any kind were so few, its consequences for the history of the human race were much more profound.

The increasing ingenuity with which successive societies have been able to exploit sources of energy other than those in human and animal muscle has given the modern world its present shape. The first uses of wind power were for propelling ships, which was how the Mediterranean came to be settled. But the development of windmills for producing mechanical power was slow. Persian innovations in the seventh century A.D. threw up the first practical designs. In medieval Europe, more efficient devices were widely used, chiefly for draining poor ground and supplying canals with water.

Water wheels first came into use in the Middle East about twenty-five centuries ago, and were chiefly used for grinding corn. The technologists of the day could choose between water wheels with a vertical axis, best suited to fast-flowing streams, and wheels with a horizontal axis. In Roman times, a single water wheel could grind 400 pounds of corn an hour, the equivalent of something like 3 horsepower. By the fourth century A.D., a corn-grinding plant at Arles, equipped with sixteen water wheels, was able to grind 7000 pounds of corn an hour. The Domesday Book, the Normans' inventory of their spoils of conquest of England, recorded 5200 water wheels in southern England, used for driving saw mills, forging hammers and ore-crushing plants as well as corn mills, and also a device at Dover for exploiting tidal energy.

The social consequences of the dependence of medieval

Europe on the water wheel were considerable, and are represented even now in the distribution of villages and towns from Italy to northern England. Reducing corn to flour is an energy-intensive process, at least by the yardstick of what a single man can accomplish. Two men working a medieval corn mill by hand and foot would expect to be able to grind 10 pounds of corn an hour. In an eight-hour day, each of them would, in other words, expect to grind 40 pounds of corn, enough to provide an adequate daily ration for some thirty people, which, in turn, implies that without the use of water wheels, something in excess of 3 per cent of the medieval population, perhaps as much as 10 per cent of the labour force, would have had to be employed in flour mills.

The economic benefit to medieval society of the water-driven corn mill was plainly substantial. For one thing, it made flour widely accessible, with all the nutritional benefits that followed. Then it released men for more useful jobs. But water wheels, corn mills and ore-crushing plants require water and capital, which, in turn, accounts for the distribution of medieval settlements along reliable water courses and for the development of the mercantile system of the Middle Ages – the complicated system by means of which the prosperous millers became not merely flour merchants but also seed merchants and, frequently, the sources of finance for agriculture.

In due course, the water wheel became the foundation of the Industrial Revolution. In Britain, the development of the new textile machinery in the middle of the eighteenth century was prompted by the imbalance within the traditional industry – the spinners could not keep up with the weavers. The spinning frames and spinning jennies, however, could only with difficulty be operated by human or animal muscles. Water wheels were, in any case, a more dependable source of power, with the result that the new cotton mills were scattered through Lancashire and the Midlands, the traditional base of the hand-spinners and weavers, in any place where a stream or a river could provide a constant source of energy. The pace of change was rapid. Between 1770 and 1800, for example, imports of cotton to Britain multiplied fifteen times, from 1600 tons a year to 24 000 tons a year. By 1851, cotton imports into Britain had multiplied by a further factor of fifteen, to 330 000 tons a year, and they doubled again in the succeeding decade.

The economic consequences were profound, as were the social consequences. Machine-made cotton was at once cheaper and of better quality – it was uniform in thickness, stronger and, if necessary, finer than even the hand-spinners of India could produce. Thus the water-powered cotton mills were an immediate benefit to the then quickly growing population of Britain in providing cheaper cloth and an important commodity of the rapidly growing export trade. They were also, of course, the spearhead of the factory system, the economic foundations for the growth of the cotton towns of Lancashire and Derbyshire, a source of the prosperity of the first manufacturing entrepreneurs, the means by which the employment of children in factories became commonplace and an important part of the legend of industrial unrest that repeatedly erupted through the first half of the nineteenth century in Britain.

The Industrial Revolution is correctly named. Few transformations of society have gone so deep. Yet the term is imprecise. Most revolutions mark an abrupt transition from one state to another. The Industrial Revolution, by contrast, was open-ended – it was a transition from stasis to a condition of repeated change. By the end of the eighteenth century, the spinning jenny was virtually obsolete, replaced by the much more productive mule. At the same time, the first power-driven looms began to replace the hand-loom weavers, for whom the closing years of the eighteenth century were a period of unprecedented prosperity. Strictly in terms of labour productivity, the gains were enormous – one young boy tending a pair of power looms early in the nineteenth century could turn out as much work as fifteen hand-loom weavers. The advantages of the new machines were plain, yet it is a considerable technical achievement that the cotton industry's population of powered looms should have multiplied forty times between 1813 and 1833, when 100 000 of them were installed in the weaving sheds of Lancashire. By the middle of the nineteenth century, there were 250 000 looms at work in Britain, the great majority still driven by water power.

The steam engine was, by comparison, a late arrival, for

reasons which are entirely explicable in the language of fuel economy. Although the first steam engines were used for emptying mine shafts of water at the beginning of the eighteenth century, their efficiency was so low that they could be used profitably only where they could use fuel that would otherwise have been wasted, chiefly the unsaleable coal from coal mines. Only in the 1780s, mainly because of the innovations of James Watt, was the steam engine able to turn more than about 1 per cent of the energy of its fuel into mechanical energy. Savery's 'atmospheric engine', widely used for pumping water from pits early in the eighteenth century, is reckoned to have used 30 pounds of coal per horsepower per hour. Watt's engines had reduced fuel consumption to 8 pounds and, after a succession of technical improvements involving higher pressures (largely an American line of development) and multiple cylinders, reciprocating steam engines used only just over a pound of coal for each horsepower in an hour. Increasing efficiency gave the steam engine its competitive edge, but from the last decade of the eighteenth century it was also clear to the new manufacturers that steam engines could provide power in previously unrealized concentration. For all the bulkiness of the Boulton and Watt beam engine, the standard samples manufactured at the rate of one every day or so at the turn of the century could each deliver 10 horse power, enough to drive several machines by means of the leather belts and pulleys which became the standard overhead furniture of the factories.

Even so, water power was not immediately driven into disuse. In the British textile industry, the capacity of the water wheels continued to increase until the late 1860s. Installed capacity of water wheels in British textile factories reached a peak of 30 000 horsepower in 1868, and only afterwards began its slow decline. But this period was one of exceedingly rapid growth for the steam engine. In textile factories alone, the installed capacity of steam engines multiplied more than four times in the two decades after the middle of the century, from 108 000 horsepower in 1850 to 478 000 horsepower in 1870. Between 1840 and 1880, the total horsepower of all the steam engines in service in Britain increased from 620 000 to 7.6 million. In the rest of Europe in the same period, the total horsepower of steam engines increased from 240 000 to 14 million. As a sign of things to come, the horsepower of working steam engines in the United States increased from 760 000 in 1840 to 14.4 million in 1880.

The steam engine left an indelible mark on the nineteenth century. The textile factories became still bigger. The iron industry was transformed, partly because steam engines made it possible to provide more powerful blasts of air than could be provided by other means – horse-driven bellows, for example. The result was that blast furnaces became larger and, as a result, more efficient. In Britain, between 1869 and the end of the century, the production of pig-iron doubled from just over 4 million tons a year in 1869 to 8.5 million tons in 1903, but during this interval, the amount of coal consumed by the iron and steel industry as a whole actually declined, from 32 million tons in 1869 to 28 million tons in 1903.

Finally, the steam engine made the railways possible. Technically, the essential step was the design of a steam engine working at higher pressures than Watt's machines, for only high-pressure engines could be made light enough in weight to provide power for vehicles. In Britain, Trevethick's 'Cornish Engine' was the starting point for the development of railway locomotives. In the United States, the inventions of Oliver Evans powerfully stimulated development. Economically, and socially the consequences were profound.

Throughout the early decades of the Industrial Revolution, the symbiotic relationship between steam and coal was plain to see. Cheap coal made steam power usable, but steam engines made coal mining productive. Even the rudimentary pumping engines of the eighteenth century made it possible to extract coal from levels in the ground that would otherwise have been flooded, with the result that the miners could follow thick seams of coal into the ground and not rely on quickly worked-out outcrops. But by the early decades of the nineteenth century, steam engines were used for other tasks as well – providing ventilation in deep mines and bringing coal to the surface of a pit.

The gathering momentum of the Industrial Revolution powerfully stimulated the use of coal, but even in Britain, accurate statistics were kept only from 1851. It is, however, known that the Newcastle Vend, which in the seventeenth and eighteenth centuries played a role for the coal industry of north-east England very much like that of the Texas Railroad Commission for the Texan petroleum industry in more recent times, increased production tenfold between 1609 and the beginning of the nineteenth century, when a total of 2.5 million tons of coal was produced. In 1854, British coal production amounted to 64.6 million tons, and had increased to 92.7 million tons a decade later, corresponding to an average annual increase of 3.5 per cent a year.

The widespread optimism of the Victorians was to some extent offset by the views of those who held that this pace of growth could not persist. The most eloquent of these was W. Stanley Jevons, whose book *The Coal Question* argued in 1865 that continued growth of coal consumption at 3.5 per cent a year was a serious threat to the commercial future of Britain. His argument was a subtle precursor of recent predictions that the present rate of growth of energy consumption cannot continue without bringing trouble in its wake.

In 1860, Britain produced 80 million tons of coal, more than the rest of the world put together. (In the same year, United States coal production amounted only to 14 million tons.) Jevons went on to project the rate of increase of coal consumption in Britain into the future, and calculated that it would have increased to 331 million tons by 1901 and to 2607 million tons by 1961. 'If our consumption of coal continues to multiply for 110 years at the same rate as hitherto, the total amount of coal consumed in the interval will be one hundred thousand million tons.' But, Jevons pointed out, the then known reserves of British coal above a depth of 4000 feet amounted to only 83 000 million tons. 'Rather more than a century of our present progress would exhaust our mines to the depth of 4000 feet, or 1500 feet deeper than our present deepest mine.' His conclusion was not that the coal reserves would at some stage be exhausted, but that:

> We cannot long maintain our present rate of increase of consumption; that we can never advance to the higher amounts of consumption supposed. But this only means that the check to our

progress must become perceptible within a century from the present time; that the cost of fuel must rise, perhaps within a lifetime, to a rate injurious to our commercial and manufacturing supremacy; and the conclusion is inevitable, that our present happy progressive condition is a thing of limited duration.

What Jevons feared most was that the cost of coal to British manufacturers would quickly exceed the cost elsewhere. He was especially fearful of the United States, where it was already clear that coal reserves were larger than in Britain or in any other European state. The lever of competition, Jevons foresaw, would be the relative prices of coal in Britain and the United States.

How did Jevons's vision of the future compare with reality? In the second half of the nineteenth century, coal consumption in Britain grew more slowly than in the first half, partly because of the increasing efficiency of steam engines, partly because the mid-century marked the point at which the pace of innovation in Britain faltered. There is room for endless argument about the reasons why the competitive advantage that Britain had acquired in the first half of the nineteenth century should have been eroded in the second, but it is hard to think that the price of coal could have been responsible. Between 1860 and 1900, the price of coal in Britain increased by only a third, and it was still cheaper by the end of the century than coal from the European mainland. In the United States, it is true, there was a steady decline of the price of coal; Pennsylvania anthracite held its price constant, but bituminous coal was being shipped from Baltimore at the end of the century at half the price obtaining in the 1850s. Yet the cost of coal was only a part of the cost of producing manufactured goods, usually (except for steel) a small part.

The Industrial Revolution in Britain faltered for other reasons. The great prosperity of the closing decades of the nineteenth century was partly an illusion, based on the cosy captive markets established in the quickly growing British Empire. Investment in innovation – the Bessemer process for making steel, for example – would have been a pointless luxury to British manufacturers with assured markets in the colonies. But the seeds of the British decline in comparison with its continental and North American competitors are more probably to be found in the particular way in which British society had evolved.

The early successes of the Industrial Revolution were in part a product of the mobility of English society. The manufacturing middle classes were recruited partly from the younger sons of the landed gentry and partly from upstart artisans. The political upheavals of the early decades of the nineteenth century served, in a curious way, to institutionalize the belief that a man must battle for his place in society. If he can better himself, that is a virtue to be applauded, but there is no particular reason why he should be helped. Is it any wonder that technical education in Britain should have lagged behind continental practice? Jevons, in other words, should have been on the lookout for other than strictly economic causes for the impending relative decline of British industry.

His prophecy was in any case vitiated by his uncharacteristically dogmatic dismissal of the possibility that there could ever be other sources of energy than coal. Wind, he said, was not a realistic alternative to coal – just the steelworks at Dowlais in South Wales would have required 1000 windmills to keep the blast furnaces and rolling mills alive. Water power, far from being a possible substitute for steam, had from the beginning of the nineteenth century in Britain been a secondary source of energy – in the mid-century, many textile mills were still being driven from mill ponds replenished by steam engines. Jevons considered and rejected the use of windmills as a means of producing electricity which might then be used to produce hydrogen gas – there was simply not enough energy in the wind to keep Britain supplied with all its needs, and in any case, hydrogen was an inconvenient fuel.

For good measure, Jevons discounted the arguments, originated by Babbage at the end of the eighteenth century, that geothermal energy might become a practicable source of industrial power; all very well, said Jevons, for those who happen to live in Iceland. Finally, 'among the residual possibilities of unforeseen events, it is just possible that sunbeams may be collected or that some source of energy now unknown may be detected. But such a discovery would simply destroy our peculiar industrial supremacy.'

In the second half of the nineteenth century, the petroleum industry was growing quickly, but Jevons did not consider it a threat to the supremacy of coal – at least in part, no doubt, because there was then no easy comparison between the costs of producing native petroleum from oil wells in the United States and petroleum from the oil shales of Scotland. He wrote:

> Petroleum has of late years become the matter of a most extensive trade . . . It is undoubtedly superior to coal for many purposes, and is capable of replacing it. But then, what is petroleum but the essence of coal, distilled from it by terrestrial or artificial heat? Its natural supply is far more limited and uncertain than that of coal, and an artificial supply can be had only by the distillation of some kind of coal at considerable cost. To extend the use of petroleum, then, is only a new way of pushing the consumption of coal. It is more likely to be an aggravation of the drain than a remedy.

The reasons why Jevons's prophecies turned out to be invalid provide not merely an explanation of a Victorian academic aberration but also a pointer to contemporary confusion about energy policy. What Jevons did not and could not know was that even a century after the invention of the first cotton-spinning water frame, the Industrial Revolution was still in its infancy. The replacement of water wheels by steam engines was such a triumph of engineering, and such a powerful catalyst of social and economic change, that it would have seemed a kind of sacrilege to look ahead to changes that would, in their ways, have still deeper consequences. Jevons and his influential supporters were saying that so long as coal remained the only source of energy, and so long as industrial output in Britain continued to increase, the time would come when the price of coal would be substantially increased and when other nations would enjoy a competive advantage. Unhappily for Britain, neither the Jevons camp nor its opponents recognized that the Industrial Revolution in which they lived was a process of continuing change and repeated transformation.

Even in mid-Victorian England, it should have been possible to know a little more about the future. The industrial history of the previous century had been a vivid demonstration of how perishable are even the most influential innovations. After all, the spinning jenny had lasted less than a quarter of a century before being overtaken by the mule. Generations of steam engines had succeeded each other more quickly than the generations of iron-masters. The land surface of Britain had been covered by a network of canals that had begun already to decline in the face of competition from the railways – perhaps the only uniquely British contribution to the techniques of the modern world. Cities in Europe and the United States had been provided with gaslight, but it was already plain that the coming of electricity would not be long delayed. Enough was already known of gas engines to know that the internal combustion engine could not be far away. Steam turbines and, eventually, gas turbines, were natural if difficult and unpredictable extensions of existing techniques. To blame the mid-Victorians for failing to foresee the heavier-than-air machine, or the nuclear power station, would be absurd, but it is remarkable that after a century of unprecedented industrial change, the Victorians could behave as if an end-point had been reached.

In reality, even the closing decades of the nineteenth century showed clearly enough that the Industrial Revolution was still in full flood with innovations such as the Bessemer process for making steel and, at the turn of the century, the development of practical methods of smelting aluminium ores. But the closing decades of the nineteenth century were marked above all by the development of new energy-transforming technologies. The steam turbine began to replace the reciprocating steam engines by the end of the century. The manufacture of electricity in bulk began slowly in the 1880s, but by the end of the century was a powerful competitor for the gas industry in domestic lighting and set in train a new flurry of excitement in railway transport. The internal combustion engine was born at the same time and was important not merely for its transformation of road transport but also because it made air transport possible. And, by the turn of the century, the petroleum industry which Jevons had scorned was based not just in the United States but also in Eastern Europe and what is now the Soviet Union. And the catalogue of innovation has, of course, been extended enormously in the succeeding decades and is still far from complete.

The Industrial Revolution, in other words, has not yet ended. It consists not of the steady growth of industry in a pattern created once and for all in the nineteenth century, but of the repeated and radical transformation of that pattern. The domestic gas industry as based on the extraction of fuel from coal came and went in just over half a century, but now, ironically, schemes for manufacturing synthetic fuel from coal are once again being talked about, especially in the United States. The radio valve, the basis of the radio industry for half a century, has now, for practical purposes, been replaced by the transistor. The passenger-carrying airship has come and gone, at least for the time being. What this implies is simply that the products and processes by which the Industrial Revolution is advanced are hardly ever permanent. In the nature of things, radical changes of technology overtake each other.

This is the background against which the upheaval of the past few years in the energy market must be assessed. Between the 1940s and the 1960s, in most industrialized nations of the West, the supremacy of coal in the Industrial Revolution was undermined by the advantages, especially the cheapness, of petroleum. For much of that time, however, it has been clear that petroleum itself could not indefinitely sustain the rapid growth of demand throughout the world. And as luck will have it, coal and petroleum are both of them among the less plentiful sources of energy likely to be available in future decades. So if there now appears to be a high chance that petroleum will be displaced from its central role in the economies of the industrialized world, it is no more surprising than the fall from favour of coal in earlier decades. If, by the end of the century, it turns out that nuclear power is the most rapidly growing alternative to the fossil fuels, that will come as no surprise either, but will rather be entirely consistent with the pattern of change since the beginning of the nineteenth century.

The impending change in the pattern of energy consumption is, in other words, neither a surprise nor a tragedy. But this does not imply that oil-consuming states need passively accept the pattern and the speed of change at present being forced on them by OPEC countries. Although conventional petroleum is relatively scarce, there is enough of it left in the ground to ensure that the changes now required could be deliberate. Indeed, as will be seen, there is a strong case for thinking that the interests of OPEC countries as well as those of the oil consumers would, in the long run, be best served if the price of oil were over the next few decades reduced to something more like $7.00 a barrel than the $10.00 or more a barrel at which it is now being sold in the Persian Gulf. One of the most curious features of the past few years has been the failure of the oil-consuming states to recognize that the inevitability of industrial change does not absolve them from responsibility for making sure that the process of change itself will be as nearly painless as possible. That, indeed, should be the objective of the wise management of the Industrial Revolution.

Unless the past few years are to seem in retrospect a sharp discontinuity in the history of the Industrial Revolution, however, it will not be enough to play politics with petroleum. Industrial communities must start to develop today the alternative sources of energy that will, in any case, be needed by the end of the century. For, as has been recognized from the beginning, both industry and the entire process of industrial change depend entirely on the availability of energy in some form or another. In the 1860s, when coal seemed the only practical source of energy, Jevons proclaimed the nineteenth century as the 'Age of Coal', and wrote this hymn to the fuel that had made the Industrial Revolution possible:

> Day by day it becomes more evident that the coal we happily possess in excellent quality and abundance is the mainspring of modern material civilization. As the source of fire, it is the source at once of mechanical motion and of chemical change. Accordingly it is the chief agent in almost every improvement or discovery in the arts which the present age brings forth. It is to us indispensable for domestic purposes, and it has of late years been found to yield a series of organic substances which puzzle us by their complexity, please us by their beautiful colours, and serve us by their various utility. And as the source especially of steam and iron, coal is all powerful . . . Coal in truth stands not beside but entirely above all other commodities. It is the material source of the energy of the country – the universal aid – the factor in

everything we do. With coal, almost any feat is possible or easy; without it, we are thrown back into the laborious poverty of early times.

In the 1930s, George Orwell made the same point, more tartly, in *The Road to Wigan Pier*:

> Our civilization is founded on coal more completely than one realizes until one stops to think about it. The machines that keep us alive, and the machines that make the machines, are all directly or indirectly dependent on coal. Practically everything we do, from eating an ice to crossing the Atlantic, and from baking a loaf to writing a novel, involves the use of coal, directly or indirectly. For all the arts of peace, coal is needed; if war breaks out, it is needed all the more. In time of revolution, the miner must go on working or the revolution will stop, for revolution as much as reaction needs coal. In order that Hitler may march the goosestep, that the Pope may denounce Bolshevism, that the cricket crowds may assemble at Lords, that the Nancy poets may scratch one another's backs, coal has got to be forthcoming.

These statements have seemed old-fashioned for at least a quarter of a century, but they still contain a kernel of truth. Whatever steps may be taken in the years ahead to economize in the use of energy, it is beyond dispute that modern industry could not be sustained without plentiful supplies of energy in bulk, for that is the only means by which the physical energies of human beings can be multiplied so as to produce the goods and services which are now valued for their own sake, but which are also the basis of the division of labour in modern society that has made possible the rapid and largely beneficent growth of social services, including health care and education.

As it turns out, in the next few decades the only practical way of providing energy in abundance, as an alternative to the fossil fuels, is by the rapid development of nuclear power. The vision of some that solar power or geothermal power might do instead is, for the time being at least, and probably forever, an illusion, but nuclear power is still far from being a familiar technology. The question which the oil-consuming states must answer soon, and should indeed have answered at least a decade ago, is whether they will now devote to the development of nuclear power the resources the enterprise needs.

2
Patterns of consumption

In the consumption of energy, the early stages of the Industrial Revolution were closely similar in Britain, the rest of Europe and North America. The new industrial enterprises sprang up where energy was easily to be had and, as a consequence, was cheap. So long as water power was the chief source of energy, industry was confined to the river valleys. When coal began to make a marked impression on the developing economies of the West, the pattern of industrial development was inevitably determined by the geography of the coal fields. The reasons were simple, and largely economic. But now, in the mature phrase of the Industrial Revolution, there is much more flexibility.

First, the traditional sources of energy have been joined by petroleum. Second, the development of means by which the energy can be converted from one form into another has liberated industrial enterprises from a strict dependence on the location of the primary fuels. Finally, the development of means of transporting energy, sometimes over great distances, has still further increased the ease with which changes may be rung on the changing pattern of industry. The gas pipelines which span North America and Eastern Europe, the high-voltage transmission lines for electricity which are everywhere to be found and the technology of ocean and pipeline transport for petroleum have transformed the moulds within which industry must grow. Greater flexibility has meant that new opportunities have been made accessible – this, for example, is the spirit in which Japan has grown to be a great industrial power. But the importance of making accurately economic choices of which sources of energy best suit which purposes has also been increased – and one of the underlying causes of the present upheaval in the supply of energy is the way in which the governments of industrialized states have frequently intervened to make rational economic choices impossible.

If Jevons's nineteenth century was the Age of Coal, the present must surely be the age of choice. It is not therefore surprising that different countries that are industrially and economically very similar have evolved distinctive patterns of energy consumption. From this it follows that the industrialized states now affected by the increased price of petroleum should rationally respond in different ways to the changing climate in which they function. What follows is an account of what the energy economies of the industrialized nations have in common and of the ways in which they differ.

The concept of energy needs prior definition. The steam engine notwithstanding, one of the triumphs of the nineteenth century was the doctrine that energy cannot be created out of nothing or made to vanish. If a quantity of coal is burned so as to turn water into steam, it is beyond dispute and, so far as the physicists are concerned, almost a matter of definition that the energy released from the fuel is numerically identical with the amounts of energy consumed in heating the water to its boiling point and then turning it into steam together with whatever heat energy may be wasted in the process. The use of the word 'energy' to describe the ways in which people keep themselves warm, drive steam engines or make electricity was meaningless until this point had been established, in the middle of the nineteenth century.

All industries are devices for turning materials of one kind into some other form – flint-stones into arrow-heads, methane into polyethylene – and all of them require energy to be effective. Some industries exist simply to transform one kind of energy into another – the energy of burning coal into the energy of electricity, for example. In some of these activities, several energy transformations may take place in sequence. In a conventional electricity plant, for example, the heat released by burning a fossil fuel is first turned into the energy of high-pressure steam, then into the mechanical energy of a swiftly rotating turbo-alternator, and finally into the energy dissipated by the electricity that the whole system produces. The amount of energy that can be won from the electricity produced is less than that originally released by the burning of the fuel, for some heat escapes from the boiler, some heat is lost from the turbine casing, large amounts of heat are made useless when the steam is condensed at the back-end of the turbine, while both the electrical generator (which in modern practice must be actively cooled by a current of hydrogen so as to carry away the energy) and the transmission lines are further thieves of the original energy of the fuel.

The principle that the total amount of energy cannot be altered leads naturally to schemes for the measurement of energy as distinct from, say, amounts of coal or oil or uranium. Indeed, there are several equally valid alternatives. One of the first units of measurement for energy, the British Thermal Unit (BThU), is now, paradoxically, used only in the United States. On mainland Europe, the traditional unit of measurement has been the calorie (not to be confused with the term dieticians use, which is 1000 times as big as the scientific unit). More recently, the unit of measurement called the joule, which began life as a unit of measurement of mechanical or electrical work, has tended to replace the calorie (1 calorie =4.2 joules). In Britain, but hardly anywhere else, energy is often measured in units called therms (1 therm=100 000 BThU, or approximately 25 million calories). The amounts of energy that can be had from different fuels naturally varies with their quality.

Some other units of measurement, such as horsepower and kilowatt, are not measures of energy but of the rate at which energy is produced, transformed or used. Power, not energy, is the word. But one horsepower or one kilowatt for some interval of time, say an hour, is an amount of energy; whence, in particular, the unit of electrical energy called the kilowatt hour. One horsepower is roughly three-quarters of a kilowatt.

The coexistence of these alternative units of measurement is in many ways an intellectual comfort. They are proof of how familiar is the notion that energy can be turned from one form into another. There are, however, pitfalls. In converting heat energy to mechanical energy, in internal combustion engines, steam engines and electricity plants burning fossil fuels or even uranium, there are unavoidable limits to what can be done. The principle is that of the Second Law of Thermodynamics, best known as the test advocated by C. P. Snow for the literacy of non-scientists. What this principle implies for the conversion of heat energy to mechanical energy is that the efficiency of conversion is always less than perfect, and is determined by the physical conditions under which the engine functions. The energy of the fuel which fails to appear as mechanical energy does not vanish into thin air, but appears instead as waste heat, as in the cooling water from a power station or the exhaust gas from a petrol engine.

For a steam engine, for example, the higher the temperature (and, in many circumstances, the pressure), the greater the efficiency with which heat energy is converted into mechanical energy. The steady increase of the efficiency of reciprocating steam engines during the nineteenth century stemmed, in part, from the steady increase of working temperature. More recently, the efficiency with which the steam turbines of electricity generating plants convert the heat of fossil fuels to electricity has been increased by similar developments. The energy that fails to appear as electricity is disposed of by whatever method is used for condensing the steam exhausted from the turbines, and the scale on which this 'waste heat' is now disposed of has given rise, in recent years, to fears of 'thermal pollution' of the environment as well as to unfounded complaints that electricity generation is a wasteful way of using natural resources. The truth is that electricity generation is more efficient than most other ways of turning the energy of fuel into mechanical energy.

Academic though these questions may seem, they have an important bearing on what is called energy policy. Just as physicists proclaim that one form of energy can be freely converted into others, so those who consume energy are often indifferent to the sources from which it comes. A domestic washing machine or an electrically driven lathe will function equally well whether the electricity comes from a conventional or a nuclear power station. This is why it can be illuminating to work out the aggregate amount of energy consumed by a national economy, and to consider how this changes in the course of time. But there are obvious limits to the ease with which these substitutions can be made. A nation's stock of motor cars cannot be converted overnight to run on coal, for example, so it follows that energy policy must be concerned not merely with the security of supply but also with the arrangements for substituting one source of energy by another.

The growth of energy consumption in the past half-century has been rapid and sustained. In the world as a whole, the consumption of energy increased from the equivalent of 1050 million tons of crude oil in 1925 to the equivalent of 3850 million tons of crude oil in 1965. Inevitably, the pace of growth of energy consumption was faster than that of the population. In 1925, energy consumption was the equivalent of 0.54 tons of crude oil for each person in the world, but by 1965 *per capita* consumption had more than doubled, to the equivalent of 1.17 tons of oil a year.

Plainly it is important to understand the means by which this growth was sustained, and to appreciate what benefits accrued from it. Some statistical pitfalls are, however, worth keeping in mind.

First, in converting figures for the consumption of different fuels into an aggregate, there are uncertainties which stem from the variation of energy content of the same kinds of fuel from different regions. The energy in a ton of coal can vary with the seam from which it has been mined, as can the energy in a ton of oil from different oil fields, and even the most meticulous statisticians cannot accurately take account of all these variations in working out the aggregate consumption of energy. In what follows, it has been assumed that 100 tons of coal is the equivalent of 68 tons of crude oil, i.e. that one ton of oil is the equivalent of one and a half tons of coal.

Second, there are problems in knowing how to count the energetic equivalent of hydroelectricity, which may be represented either as the energy of the electricity produced or as the energy of the fossil fuel that would be needed to produce the same amount of electricity. Where aggregate energy consumption is concerned, the second is the more logical course to follow. (In this case – allowing for the fact that hydroelectricity generators are usually kept operating almost continuously, and for the efficiency with which conventional power plants convert the energy of fossil fuels into electricity – one kilowatt of installed capacity in a hydroelectric plant is the equivalent in one year of between 2.0 and 2.5 tons of oil.)

It is also important to know that the pace and pattern of the growth of energy consumption has been distorted over the past few years by several developments, especially the increase of oil prices, so there is much to be said for regarding the period up to 1965–70 as that which shows the normal development of the energy industries of the world.

The growth of energy consumption in the past few decades has been met by changes in the pattern of supply every bit as dramatic as those of the nineteenth century. Between 1925 and 1965, for instance, the consumption of coal increased by a mere 1.6 per cent a year – it failed to double. By the end of this period, coal accounted for just under 40 per cent of all energy consumed, compared with roughly 80 per cent in 1925. The middle decades of this century were dominated by the rapid increase of oil consumption and, to a lesser extent, by growing importance of natural gas. For forty years from 1925, the world's consumption of petroleum liquids increased at an average annual rate of 6.2 per cent, so that even by 1965 oil accounted for almost as much energy as that supplied by coal, a balance since tilted further in favour of oil. Natural gas consumption has grown more rapidly still, at an annual rate of 7.6 per cent between 1925 and 1965, and at the end of the period was the equivalent of more than 600 million tons of crude oil a year. The middle third of the twentieth century has plainly been as much an age of oil as the nineteenth century was an age of coal.

Aggregate figures for energy consumption, like aggregate figures for the world's use of telephones or television sets, conceal important variations in the pattern of consumption and its change. It is no surprise that the more industrialized or more prosperous countries use more energy than others less fortunate, for the relationship between energy and industry is self-evident. What is surprising is the recent fashion in the United States for breast-beating which takes the form of the complaint, 'Why should 6 per cent of the world's population consume a third of the world's energy?' The most immediate answer is that the United States also produces roughly a third of the world's energy. In any case, the implicit self-criticism in the question is not especially convincing so long as it is innocent of proposals for abandoning some forms of industrial activity.

The truth is that in the years ahead the pattern of the world's energy consumption can be influenced significantly only by the policies adopted by the principal energy-consuming regions, chiefly North America, Western Europe, Japan and the Soviet Union with its East European partners. In 1965, these regions consumed just over 80 per cent of the world's energy, and although this percentage has been falling steadily in the decades since the Second World War, the principal industrial countries of the world still consume more than three-quarters of the energy. The impoverishment of the rest of the world is evident, but it is absurd to ask that this imbalance in energy consumption should be corrected by a unilateral reduction of energy consumption in the West. Energy consumption is not merely the foundation of industry and prosperity, but also a consequence of economic consumption. The only means by which poorer countries will be able to consume a fairer share of whatever energy is available is by economic development, and it is proper to complain that so far this has been inadequately fostered by the half-hearted efforts of the prosperous nations during the decades since the Second World War.

The link between a nation's energy consumption and its prosperity is now well documented, most simply by the relationship between energy consumption and Gross National Product (GNP). The simple rule is that the greater a nation's GNP, the greater is its energy consumption. There seems to be an almost mathematical relationship between prosperity (at least as measured by GNP) and energy consumption, but the variations are nevertheless significant. Both Canada and South Africa, for example, consume more energy than might have been expected because both of them are heavily engaged in the metal-extraction industries, which are notoriously heavy consumers of energy, often as electricity. The countries of Western Europe, differing markedly both in energy consumption and prosperity, are relatively modest energy consumers, a measure partly of their relatively small involvement in metal extraction and partly of the way in which relatively high energy costs have encouraged energy economy. By contrast, the countries of

Eastern Europe use more energy than might be expected, which is most probably a sign of the speed with which they have changed in the past half-century and of the way in which market forces are blunted by their political systems, with the result that there are perpetuated into the present wasteful practices that have elsewhere been weeded out.

Although *per capita* GNP and energy consumption are closely linked, the experience of the past half-century has also shown that increasing prosperity often brings a decrease of the intensity with which energy is used. Between 1925 and 1965, for example, in both the United States and the United Kingdom, the amount of energy consumed for each dollar's worth of GNP declined slowly but steadily by almost a third. In 1925, in both countries, each dollar's worth of GNP required the consumption of the equivalent of roughly three litres of oil (the equivalent of two-thirds of an imperial gallon, or nearly four-fifths of a US gallon) to just over two litres of oil for each dollar of GNP.

Much the same tendency has been apparent in other countries with mature industrial economies, in which economic development usually entails the growth of sophisticated industries (electronics, for example). Even so, it is remarkable that in countries such as Japan energy consumption per unit of GNP is only three-fifths of that in heavily industrialized countries such as Britain and the United States, no doubt because of the way in which a shortage of indigenous energy supplies, and the consequent higher cost of energy, stimulates energy economy. Elsewhere, as in predominantly agricultural countries, it is easier to understand why the amount of energy corresponding to each unit of GNP should be lower than in highly industrialized countries, while deliberate changes in the pattern of a country's economy naturally bring changes in the pattern of its energy consumption. The most vivid illustration is the growth of energy consumption in the Soviet Union since 1925, where the amount of energy consumed for each unit of GNP multiplied six times in forty years.

Another proof of the relationship between economic growth and energy consumption is to be found in the statistics compiled to show the relationship between the rate of growth of energy consumption in different countries and the rate of economic growth (measured by GNP). Between 1950 and 1965, this ratio ranged from roughly 0.75 in countries such as Britain, West Germany and the United States (which implies that energy consumption was growing less quickly than GNP), to more than 2.0 for countries such as Argentina, mainland China and Bulgaria.

The implication is that individual nations in the years ahead should have ample opportunity to adjust their pattern of economic development to the changing pattern of energy supply without forsaking economic growth. But there are two important reservations. First, the history of the past few years shows clearly enough that important changes in the intensity of energy consumption do not happen quickly. Secondly, a simple consideration of the relationship between energy consumption and GNP cannot by itself be a useful guide to policy. One obvious difficulty, for example, is that energy consumption is, in part, a consequence of personal consumption, which is in any case a varying fraction of the GNP and which may be made to vary by the policies which governments pursue.

The experience of the United States between 1965 and 1970 is a curious and still not entirely explicable exception to the rule that economic growth tends to reduce the intensity of energy consumption. During that period, while GNP was increasing at 3.1 per cent a year, total energy consumption increased by 5.0 per cent a year. The President's Council of Economic Advisors reported in 1972 that the sudden acceleration in the growth of energy consumption was a sign of fundamental changes in the energy market, predicted that during the 1970s the growth of energy consumption and of GNP would once more be in balance, but argued that the uncertainties were too great for accurate predictions. The contributory causes of the rapid growth of energy consumption in the United States in the late 1960s included the expansion of non-energy uses of petroleum (such as feedstocks for chemical plants), the tapering off of the steady improvement of the efficiency of electricity power stations which had been in use for half a century and the increasing use of air conditioning and electric heating by American domestic consumers. Although the speed of this increase of energy consumption is still not properly understood, its practical consequences was to sharpen the dependence of the United States on imported oil and to make its appearance in the Middle East market more abrupt.

The acceleration of the growth of energy consumption in the United States in the late 1960s is yet another reminder that a proper understanding of a country's energy economy, and an accurate assessment of its flexibility, requires a more detailed knowledge of the activities on which energy is expended and the sources from which it is derived. In spite of the generality of the historical link between economic growth and energy consumption, different countries vary enormously.

In the sources of energy, as in other ways, the United States is an extreme case. To be sure, it is not surprising that petroleum should account for something like 44 per cent of all energy consumption (only a little above the world average), for much of the rapid growth of the United States in the past half-century has been sustained by the indigenous reserves of Texas, Louisiana and California. In 1971, the United States consumed 755 million tons of liquid petroleum. But the United States differs from other industrialized nations in its substantial use of natural gas, which in 1971 accounted for roughly a third of all energy consumption. This is partly a measure of the geographical availability of natural gas, which can be transported easily and cheaply only by pipeline, and partly a consequence of the Federal Government's policies, dating back to the early 1960s but amended early in 1974, of ceilings on the price of natural gas. The result has been a rapid stimulation of the use of natural gas in the past three decades, and even the uneconomic use of this valuable fuel for generating electricity. In bulk, coal remains an important source of energy in the United States, and its use increased steadily from the early 1960s, chiefly as a means of meeting the growing demand for electricity. Even so, in 1971, coal accounted for less than a fifth of total energy consumption in the United States.

The industrialized nations of Western Europe have a distinctively different pattern of energy consumption. Since 1960, coal, the basis of the Industrial Revolution, has declined steadily both numerically and as a proportion of the total energy consumed. Indigenous coal production in the OECD countries declined by 26 per cent between 1960 and 1970, chiefly because of the deliberate decision of the European Iron and Steel Community in the 1950s that European coal was needlessly more expensive than imported oil from the Middle East. Natural gas has been an important source of energy and of foreign exchange to the Netherlands since the mid-1950s, and will undoubtedly be increasingly important as a source of energy for the United Kingdom throughout the 1970s and 1980s, but it contributed only 7 per cent of Europe's energy requirements in 1970. During the 1960s, oil was the linchpin of the European fuel economy, and the bulk of what Europe used came from abroad, principally the Middle East. By 1970, imported oil accounted for 58 per cent of Europe's total energy consumption (with indigenous production, chiefly in West Germany, representing just 2 per cent of the total).

But some countries are less equal than others. By 1970, Denmark was dependent on imported oil to the extent of 92 per cent of its need of energy, and moreover had the bad luck to fall foul, in October 1973, of the Arab embargo on shipments to 'unfriendly' nations. No wonder that the district heating schemes allowed Danish people the comfort of warm water only in alternate weeks during the closing months of 1973.

The developing countries are a different case again. Europe and Japan (and Japan resembles Denmark in its reliance on imported energy) have at least the chance of being able to create (or borrow) the capital needed to make energy in quite different ways. The developing countries which have no indigenous oil of their own, but which have been persuaded quite logically by the European experience in the 1960s that energy is best had cheaply by buying oil from the Middle East, are less well placed. Unhappily, they are for the immediate future trapped by the extent to which their economic prospects are linked with the speed with which their population happens to be growing. If, by geographical accident (as in Nigeria or Indonesia), there is plenty of oil, then the chances are that the pace of beneficent development will be increased. If, for the time being, there is hardly any oil worth speaking of, as in the archetypal developing country India, then the development process will be postponed, perhaps for a decade, perhaps for longer.

The uses which different countries make of the energy they consume are as variable as the sources from which their raw materials are derived, but there is one common thread of the greatest importance for the years ahead. Throughout this century, and throughout the world, electricity consumption has been growing more quickly than the consumption of energy as a whole. Between 1925 and 1965, the consumption of electricity in the world as a whole multiplied eighteen times, corresponding to an average rate of growth of more than 7 per cent a year, while the *per capita* consumption of electricity has multiplied by ten, from roughly 100 kilowatt hours a year in 1925 to more than 1000 kilowatt hours a year in 1965.

The general rules which relate GNP to energy consumption as a whole apply also to electricity consumption, but more sharply. The richest countries may consume the most electricity, but the rate of growth of electricity consumption tends to be less rapid in the most advanced communities. This is not, however, inconsistent with a vast spread of electricity consumption, ranging from 7500 kilowatt hours per person per year in Canada to 6500 kilowatt hours per person in Sweden, 3750 kilowatt hours in Britain and less than 100 kilowatt hours per person in India.

The reasons why the consumption of electricity is everywhere growing at this speed are not hard to discover. From the beginning of the century, it has been plain that electricity is more efficient, cheaper and more convenient than traditional ways of using energy. The conquest of the lighting market was on the cards as soon as the first substantial electricity generators had been built. Electric motors have inexorably replaced steam-driven machinery in textile mills and steel mills. And it turns out to be a much more efficient use of energy (and much cheaper) to heat a billet of steel (or any other metal) in an electric induction furnace than in one of the old-fashioned gas-heated furnaces of the 1930s. But the domestic consumption of electricity has also grown steadily, and as rapidly as the use of electricity in industry. The refrigerator and deep-freeze set have been followed by the electric toothbrush and the electric carving knife. The US Federal Power Commission estimates that a simple air conditioner installed in an office window uses nearly 1400 kilowatt hours a year, the equivalent of between two and three barrels of oil, but that an electric toothbrush uses only 5 kilowatt hours a year. In the United States, the widespread adoption of air-conditioning in the past three decades has brought about a dramatic change in the pattern of operation of many utility companies: in the old days, the greatest demand on the system used to come in the winter; now the summer is the season in which the power stations are working flat out. In the United States, industrial users account for roughly 40 per cent of all the electricity generated, and domestic users for a quarter. The remainder is consumed by farmers, transport undertakings and municipal authorities or is lost in transmission or otherwise unaccounted for – between 8 and 9 per cent of the total electricity generated in the United States, as in the United Kingdom.

In such circumstances, it is not surprising that there has been some futuristic talk of the impending arrival of the 'all-electric economy', the situation in which the bulk of a country's energy supplies would be delivered in the form of electricity. However, the rapid growth of the consumption of energy in the form of electricity had been rapid enough in the 1950s and 1960s, before the large-scale development of nuclear power was seriously considered, to provide these speculations with a solid foundation. Even if fossil fuels of various kinds were to continue indefinitely as the chief sources of energy, the chances are that the proportion converted into electricity would have increased for several decades. But it is also true that even in advanced communities the proportion of the energy consumed which is converted into electricity is still only a small if substantial fraction.

Projections of electricity consumption into the future are critically dependent on the cost of electricity to consumers as well as on the persistence of an economic climate in which manufacturers and domestic consumers have the capital to buy the machinery that ultimately makes a demand on the capacity of the electricity utilities. Such projections are necessarily hazardous, but the fact remains that even strictly statistical projections into the future in the United States foreshadow a state of affairs in which, by the end of the century, roughly half the total energy consumption would be used for making electricity. To the extent that the problems of the petroleum industry in the past few years are certain to have provided a further incentive for the development of nuclear power, the coming of the all (or half) electric economy has been accelerated. Elsewhere, in Western Europe, the same considerations will apply, but with an appropriate time lag.

Despite the close link between economic growth (as measured by GNP) and the scale of energy consumption, there persist sharp and significant differences in the uses different countries make of energy. The United States is again an extreme example. Throughout the 1960s, no less than a quarter of all the energy consumed (and more than a half of all the oil) was used in transportation of one kind or another. By contrast, the proportion of energy used in people's houses was less than a fifth. Towards the end of this period, the substantially smaller consumption of energy in Western Europe than in the United States was accounted for by the more modest use of energy in transport. In 1970, in the United States, transport accounted for nearly 42 per cent of the total output of refined petroleum, itself the equivalent of 3.3 tons per head of population. Without this lavish use of petroleum products in transport of various kinds, the United States would have used 1.96 tons of petroleum per head per year, or not very much more than the average non-transport use of 1.58 tons per head in Japan. Indeed, the question has been repeatedly asked whether much of the United States' petroleum problem might not vanish if less energy were used for moving people and goods from one place to another. The answer is not merely that geography is against any such radical recasting of the energy economy of North America, but, more importantly, that US governments are politically constrained from seeking too rapid and radical a change in the pattern of energy consumption.

One surprising if not unexpected development in the past few decades has been the increasing use of energy in agriculture, especially (again) in the United States. Both by the intensive use of fertilizers in modern agriculture and the development of methods for the irrigation of crops, American farmers are increasingly supplementing the natural energy of the sun, on which all crops depend, by artificial energy, for the most part derived from fossil fuels. In the wake of the increase of petroleum prices, it has frequently been argued that the next few years will see a regression of North American agriculture to the practices of the 1930s, for which reason it is important that the quantities of energy involved should be counted accurately. One supposedly authoritative balance sheet for the energy content of an average acre of maize corn (which takes account not merely of the energy needed to make the fertilizers and drive the tractors, but also of the electricity used to keep farmers' television tubes alight), has reckoned that growing it needs energy the equivalent of 2.15 barrels of oil, itself nearly a sixth of the energy content of the grain. Yet this is not an argument against energy-intensive agriculture – rather the opposite. The intensification of North American agriculture has made it possible to increase the average yield of corn 2.5 times. In other words, for each 2.15 barrels of oil invested in intensive agriculture, it is possible to win extra corn whose energy content is the equivalent of 7.5 barrels of oil.

None of the other variations between countries in the use of energy is nearly as striking as the American predilection for burning petroleum products in internal combustion engines and jet engines. The growth of the extraction industries in countries such as Canada, Australia and South Africa has had an important influence on the growth of consumption in the past two decades, as has the concentration of petrochemicals manufacture on the growth of petroleum consumption in countries such as the Netherlands and Japan. For the most part, however, the growth of energy consumption in recent decades has had consequences similar to those of the growth of the energy industries a century ago. A host of new technologies has been made possible. The comfort and well-being (or, at least, the sense of well-being) among the populations of industrialized societies has been enhanced. And whatever happens to the price of oil, there is no sign that the process will be halted.

So has it been worthwhile, the invention of the axe, the wheel, the reciprocating steam engine and the nuclear power station? Is the cheerful belief of the Victorians in the social beneficence of energy consumption sustainable in contemporary society? To pretend that the increase of GNP which energy consumption has made possible represents a commensurate increase in human happiness would, of course, be absurd. But it would be equally absurd to suggest that the consumer goods on which the people of industrialized societies now spend a large part of their incomes are irrelevant to their contentment. To stay at home and be reasonably warm is something to be grateful for. To travel and thus enjoy the sense of freedom that even in Victorian times was the privilege of the rich is something else. And as good luck will have it, there is enough flexibility in the pattern of energy consumption, at least in developed countries, to ensure that the increasing supply of these and other goods will not be permanently interrupted by the upheaval of the petroleum market in the early 1970s.

3
New lodes to work

The most conspicuous irony of the past few years is that the increase of oil prices has bcen called a crisis when alternative sources of energy are at once more diverse and potentially more plentiful than ever before. Not only are there new sources of energy such as nuclear power, but also new means of winning usable fuels from traditional materials such as coal. So the question is not whether modern industry can be supplied with the energy it needs, but how can governments best foster the most economic blend of energy technologies? The alarm engendered in the West by the increase of oil prices is a sign that even in these supposedly rational times, governments are tempted to translate the inevitability of change and the need for choice into problems of the avoidance of cataclysm.

For decades, not just in the past few years, energy policies pursued by governments have been made contradictory by a widespread failure to appreciate that what is technologically possible must be qualified by an appraisal of what is economically desirable. Thus have the British government and many of its electors been sustained by the prospect that, in the 1980s, oil production from the North Sea may be more than sufficient to meet the British demand for petroleum. But in reality at least a part of this oil is likely to be expensive even by comparison with the cost of OPEC oil, which implies that North Sea oil will be economically advantageous for Britain only if there are further increases of the price of oil on the international markets. In short, the prospects that the North Sea may contain very large quantities of oil is not, in itself, an assurance that Britain has no further need to explore alternative sources of energy.

But why, if energy is indispensable, should its cost be important? And what should be the role of governments in the regulation of costs?

It has recently become fashionable to deride economists for their role in the management of public policy. Whatever the rights and wrongs – and they are mostly wrongs – of the arguments of the environmentalists that economic growth (as measured by GNP) is inherently undesirable, the concept of the cost of producing a commodity is indispensable not merely in commerce but in any rational discussion of public policies on energy. This concept, which dates back at least to the days of Adam Smith, is simple. If a man makes a mousetrap or a motor car, the cost of production is the cost of his labour and the raw materials he uses together with some allowance for his use of capital equipment and for the cost of raising the finance to keep himself in business. The production cost is therefore a true measure of the amount of economic resources needed to produce a single item. A manufacturer of mousetraps or motor cars can set out to reduce his production cost in several ways. He can make fuller use of his capital equipment, or he may find that substituting machines for men is more economical, or he may be able to make his manufacturing process more productive or to use cheaper raw materials.

In the energy industry, exactly the same considerations apply. The cost of producing coal from underground mines is partly determined by the cost of miners' wages and partly by the initial cost of the enterprise, the hole in the ground and the machinery installed in it. If the calculations are made accurately, the average production cost of a ton of coal is a true measure of the economic resources consumed in the process. And if coal or some other fuel can be produced more cheaply by some alternative process, the difference of production costs is a true measure of the economic resources thus liberated for some other use.

One of the characteristics of the energy-producing industries is that there is rarely a single rule for determining the production cost of a particular fuel. The cost of producing coal from shallow mines is less than that from deep mines, the cost of oil from the productive reservoirs of the Middle East is less than that of oil from Texas or the North Sea. If a community has the freedom to choose, it is an assault on reasoned self-interest to produce energy from high-cost sources when cheaper energy is available. If the choice is between importing petroleum at a high price and producing an alternative source of energy from indigenous reserves, the question that must be answered is whether the capital and the labour necessary to exploit the indigenous reserves might not be invested in some other enterprise that would bring important gains to the economy. All this should be the basis of the energy policies which nation states pursue. Too often, in the past, hard-headed calculations of, say, the cost of producing coal have been obscured by short-term considerations of the social consequences of running down the coal-mining industry.

Production cost does not however unambiguously determine the price at which a fuel is or should be sold, which is where the influence of governments is all-important. In nineteenth-century Britain, dependent simply on coal when mine owners were freely in competition with each other, the price of coal in different parts of the country was determined by the cost of production at the most expensive coal mine after allowing for the fact that the coal from different mines would be affected differently by transport costs. It follows, of course, that mine owners who were fortunate enough to be operating low-cost mines made larger profits than others. But, for the community as a whole, the best use is made of economic resources when prices are for practical purposes set by the cost of production from the most expensive mine. To say this is merely to put somewhat starkly the doctrine that the most efficient use of resources is that which obtains when the prices charged for different products are those fixed by the cost of production from the most expensive source.

Governments, alas, are often unwilling to accept the logic of this argument, even when it is properly qualified by the consideration of the folly of too strict an application of marginal pricing by, say, neglecting the social costs of putting coal mines out of business. In the United States, for instance, the fact that low-cost producers will make large profits in a competitive economy has led to the development of elaborate arrangements for controlling the prices at which natural gas and petroleum can be sold, with the result that these commodities have frequently been marketed at prices way below those that would have been charged for them in an open market. The waste of economic resources that has followed consists of the overstimulation of the market for natural gas. In Britain, exactly similar consequences have followed from the attempts by successive British governments to shield consumers of gas, coal and electricity from the full effect of increasing costs, making nonsense of the half-hearted attempt in December 1974 to exhort the British people to use energy more economically.

The argument that the prices charged for different forms of energy should be fixed by the yardstick of the most expensive source does not, of course, prejudge the question of who should fix the prices. Theoretically, at least, there is no reason why state-regulated enterprises should not follow the rules that would be determined by free market competition. Government intervention need not be inconsistent with the economists' contempt of the proper relation between production cost and prices, although, in practice, governments appear consistently to be tempted to rig the market. This is perhaps the strongest reason for allowing prices to be determined by commercial competition – governments are thereby defended from temptation.

The recent upheaval of the petroleum market has partly been caused by serious distortions through the actions of oil consumers and oil producers, but there are also geological reasons. Petroleum is intrinsically scarce compared with other fossil fuels like coal or oil shale. The petroleum hydrocarbons, oil and natural gas, have, like the other fossil fuels, been formed from organic material which has at some stage been buried in or with the sedimentary rocks. The coal seams found in many parts of the world are relics of the luxuriant forests of the Carboniferous period of 220 million years ago, and some coal seams have generated methane which has accumulated to form natural gas in overlying geological formations.

Since the emergence of life on the surface of the earth 500 to 600 million years ago, substantial quantities of finely disseminated organic material have been incorporated into all sedimentary rocks. It has been estimated that something like

1000 million tons of organic matter is at present being corporated each year into newly formed sediments, the muds and clays that will eventually be converted into rocks. And all the sedimentary rocks formed after the emergence of life are found to contain measurable quantities of organic matter, rarely less than 0.1 per cent by weight, often more than 1 per cent. Yet nobody would seriously consider mining these ancient sedimentary rocks for the pint or so of oil that might be won from each ton. The petroleum industry owes its existence to the natural processes by means of which organic matter in deeply buried sedimentary rocks has, in exceptional circumstances, been converted into simpler hydrocarbons such as those of crude oil or natural gas by the effects of high temperature and pressure.

The formation of a commercially valuable oil or natural-gas reservoir is always the result of a concatenation of geological accidents. First, there must be underlying rocks to provide a source of organic material. Secondly, there must be a geological structure that will halt the upward migration of the hydrocarbons – a layer of salt, some other impervious rock or even a geological fault. So exceptional are these circumstances that the hydrocarbons now trapped in underground petroleum reservoirs are only a fraction, possibly a small fraction, of the materials which have in the past 500 million years migrated upwards from buried organic sediments to be dispersed on the surface of the earth and in the atmosphere.

Where all the circumstances are right, the consequences are remarkable. The great Ghawar field in Saudi Arabia, an elliptical geological structure 135 miles long, consists of four layers of limestone laid down roughly 150 million years ago and separated from each other and capped on top by layers of salts formed by the evaporation of the shallow seas which once covered this part of the Arabian Peninsula. The Ghawar field is thought originally to have contained at least 6000 million tons of recoverable crude oil, and is at present yielding more than 60 million tons of oil a year.

The most valuable oil reservoirs are structures in which the rocks are porous, enabling them to contain large quantities of oil or gas (up to 30 per cent by volume), and also thick, so that a hole drilled in the ground can collect hydrocarbons rapidly.

The world's largest oil field, the Burgan field in Kuwait, is an underground layer of sandstone 1300 feet thick, conservatively estimated to have contained 10 000 million tons of recoverable crude, more than all the oil fields of the North Sea put together. Experience has shown that geologically older petroleum is chemically simpler (and thus less dense) and more likely to be free from sulphur. The deeper reservoirs are more likely to contain natural gas than liquid petroleum, but the separation between the two materials is rarely sharp – all petroleum reservoirs contain natural gas, sometimes dissolved in the crude oil, and most natural gas reservoirs yield a certain amount of liquid petroleum at the surface. Although the petroleum engineers are proud of how they have been able to drill to depths of four miles or so, the geologists and the oil company accountants are impressed that two-thirds of the liquid petroleum now being extracted comes from wells less than 1500 feet deep.

The exceptional circumstances in which oil reservoirs are formed has an important bearing on the economic cost of petroleum. A survey carried out in 1970 estimated that 30 000 oil fields were then known, of which only 187 had originally been 'giant' fields from each of which more than 500 million barrels (70 million tons) of oil might be extracted. Three-quarters of all the crude oil extracted by 1970 was thought to have come from giant fields. Similar rules apply to deposits of natural gas. A mere 1 per cent of all the reservoirs of oil and natural gas known in 1970 were then calculated to contain 65 per cent of the world's known reserves. The two largest oil fields in the Middle East, Burgan and Ghawar, are reckoned to have contained 15 per cent of the crude oil discovered by the end of 1970.

Inevitably, the larger oil fields can be found and worked most cheaply. Not only is the cost of discovery (geological speculation followed by test drilling) spread over a larger quantity of oil or natural gas, but the cost of development (sinking production wells and building well-head machinery) is likely to be smaller. Much depends on the rate at which a production well will yield oil or natural gas, which can vary enormously from one reservoir to another. Saudi Arabian oil wells average 7000 barrels of oil a day (the equivalent of 1000 tons), with the best of them more than twice as productive. By contrast, in the United States, the capacity of producing wells is typically between 50 and 100 barrels a day (250 to 500 tons a year), a somewhat artificial situation created by persistent over-drilling.

The best measure of the capital cost of producing petroleum is the amount of money spent in increasing production capacity by one barrel of oil each day. In the Middle East in the 1960s, the cost of extra capacity was on average between $100 and $150 for each daily barrel of capacity. (One barrel of oil a day is roughly equivalent to 50 tons a year). In the United States, by comparison, the cost of increasing productive capacity by one barrel a day is on average more than ten times as much – something in excess of $2000. In the North Sea, the costs of developing new petroleum production capacity are still greater – at the end of 1973, an official of British Petroleum quoted costs of £1200, close on $3000 for each barrel per day of productive capacity even in the southern part of the North Sea, where the technical difficulties are most easily grappled with. In the deep waters off the Shetland coast, development costs may turn out to be as much as $10 000 for each daily barrel of productive capacity.

Eventually, the money spent on the development of an oil field must be recovered in the price at which the oil it produces is sold, and there is no simple arithmetical rule to decide how capital cost contributes to the cost of oil produced. The rate at which an oil well is exploited, the riskiness of the venture and the cost of financing the enterprise all influence the calculations which oil producers make. But the capital cost of development in the Middle East is reckoned to add $0.05 to $0.07 to the cost of a barrel of oil (less than $0.50 a ton). In countries such as Libya and Indonesia, capital costs may be twice as large ($0.15 a barrel or $1.00 a ton). Experience in Venezuela suggests that capital investments may add $0.30 a barrel ($2.00 a ton) to the cost of oil. In the United States, the capital component in the cost of oil could probably average $1.25 or more a barrel ($9.00 to $10.00 a ton) if it were not that the situation is confused by the system of depletion allowances and other distortions of the taxation system. In even the cheapest parts of the North Sea, exploration and development is likely to account for at least $2.00 a barrel ($15 or £6 a ton) of the cost of producing oil; in the more inaccessible oil fields, capital costs may be $8.00 a barrel or more – not very different from the cost of OPEC oil in 1974, even before allowance has been made for the operating costs. Oilmen are fond of emphasizing the physical differences in the crude from different oil fields, but much the most significant difference between them is the enormous variation of cost. And oil from the Middle East is outstanding for its cheapness.

Low development cost usually also implies low operating cost. At the beginning of the exploitation of an oil field, it may be sufficient to drill a hole in the ground and let the oil flow to the surface under the influence of subterranean pressure. If long pipelines and elaborate handling facilities can be dispensed with, the cost of producing a barrel of oil is inversely related to the productivity of the wells. The faster the oil flows, the less onerous is the modest labour cost needed to superintend the operation. Later in the lifetime of an oil reservoir, the underground pressure may decrease, in which case it may be necessary to spend money on mechanical pumping or, more probably, on the maintenance of the underground pressure by reinjecting natural gas or even brine (both of which are usually produced as by-products). Secondary recovery techniques such as these are usually necessary if more than about a third of the oil in a reservoir is eventually to be extracted, but they are also increasingly regarded as means by which the natural gas always dissolved in crude oil ('associated gas' as it is called) can be kept until wanted and not simply burned.

In the productive oil fields of the Middle East, production costs are less than $0.05 a barrel in Kuwait and just over $0.05 a barrel in Iran. Libyan oil is more expensive, perhaps as much as $0.08 and $0.09 a barrel. The typical allowance in the agreements with the oil companies operating in the Middle East is only $0.15 a barrel, including both the cost of production and the amortization of the development cost. Elsewhere, as in the United States, where as little as four or five barrels of oil may be pumped each day from the multitude of stripper wells, production costs may exceed $1.50 a barrel, although the average cost of production from the more substantial wells which exist to produce oil and not simply to take advantage of the tax allowances are probably between $0.10 and $0.50 per barrel. These calculations are in good accord with the market price of American crude oil, which declined slowly throughout the 1960s to just above $3.00 a barrel by 1970.

This enormous variation in the cost of oil from different oil fields is a simple consequence of the underlying geological diversity. If at any time in the past quarter of a century there had been a free international market in crude oil, the Middle East would have driven most other oil-producing areas out of business, at least for a time. Even by the late 1960s, the 'take' of the oil producing governments, the standard royalty payment and the notional share of oil company profits calculated from the posted price of crude oil was less than $1.00 a barrel, while the cost of tanker freight to the markets of Western Europe, Japan and even the eastern seaboard of North America would almost certainly have been about $0.50, except at exceptional times, as after the Suez crisis in 1967.

The cost of developing and exploiting natural gas fields tends consistently to be less than that of producing crude oil, at least when the energy content of the two fuels is compared. Gas will more easily run up a pipeline, and the well-head equipment is less elaborate. While the delivered cost of domestic natural gas is increased by the pipeline charges and other distribution costs, ultimate retail prices have been lower than the cost of alternative fuels based on crude oil. This is the origin of one of the wounds the United States has inflicted on itself in the past two decades – demand for natural gas has been unnaturally stimulated, while the exploitation of smaller gas fields has been held back. And the same errors have been repeated, perhaps less flagrantly, in both the Netherlands and the United Kingdom.

The huge variation of the cost of producing oil from different oil fields has determined the pattern of the oil industry in recent decades. If the price of oil is now permanently increased, it becomes important to know where future oil fields will be found, and how much oil they will contain. One difficulty is that estimates of the yield of oil and gas from known reservoirs are inevitably complicated by commercial as well as geological uncertainties. Even when the amount of crude oil in an underground reservoir has been accurately determined by drilling, the owner's estimate of when it will cease to be profitable to spend money on techniques for secondary recovery must depend on his estimate of how the price obtainable for crude oil is going to change. At any time, the exploitation of some of the smaller underground reservoirs will be uneconomic – the cost of drilling even a single well may not be matched by the revenue obtainable at current prices. Especially where the cost of development is high, as in deep-water offshore areas such as the North Sea and other offshore oil and gas fields, it is inevitable that even quite large reservoirs will always be left unexploited.

Since 1960, it has also become apparent that new oil reservoirs have been more difficult to find. In the late 1940s, in the United States south of the 48th Parallel, each foot of exploratory drilling found on the average 400 barrels of oil, but only 70 barrels (for each foot of exploratory well) by 1970. Offshore, and even in Alaska south of the prolific North Slope, the benefits of exploratory drilling were much greater; in 1970, each foot of exploratory drilling found more than 700 barrels of oil – a figure which is a measure of how offshore technology has brought previously inaccessible oil fields within reach, but which is also slightly misleading, for offshore and Alaskan oil will be even more costly than that from the established oil fields of Texas and California. Although interest in offshore exploration has increased rapidly since the beginning of 1974, it still accounts for less than 10 per cent of the exploratory drilling in the United States.

Historically, the total amount of exploratory drilling for oil in the United States has until recently decreased. The reasons for this decline are far from clear. One possibility is that the major oil companies concentrated their exploration outside the United States because they had good reason to believe that exploration would be more profitable than in the most thoroughly worked-over group of the world's oil fields. No doubt this incentive to look overseas was strengthened when, towards the end of the 1960s, it became apparent that the isolation of the United States from the international oil market could not persist. But restrictions on the licensing of offshore areas, prompted by environmental concern, especially after an oil spillage in the Santa Barbara Channel off the coast of California in 1969, physically restricted their scope for activity. Drilling for gas declined during the same period, but for different reasons: through much of the 1960s it was less than fully profitable because of the way in which the price of natural gas was controlled.

The experience of recent decades is the basis for the attempts made to estimate the total amount of oil that will in future be discovered in the United States. In the nature of things, this is an exercise that has much in common with attempts to estimate the incidence of undetected crime, and the estimates that result inevitably differ. One of the most conservative of the reputable estimates is that of R. King Hubbert, who in 1969 estimated ultimate total of discoverable oil beneath the United States (offshore as well as beneath dry land) as 85 000 million tons, of which between 30 per cent and 50 per cent would eventually be extracted. Nearly 70 per cent of this, roughly 60 000 million tons, has already been discovered. In December 1972, the American Petroleum Institute produced a more cheerful estimate of reserves amounting to 115 000 million tons, 55 per cent of which had already been discovered. On this estimate, offshore areas and Alaska would, between them, account for 30 per cent of the known reserves of oil in the United States, but probably for a smaller proportion of the amount of oil actually recovered because of the economic disadvantages of exploiting marginal oil fields. Then, in 1973, the US Geological Survey produced a still more bullish estimate, that 270 000 million tons of oil would eventually be found beneath the United States and its continental shelf, and that as much as a half of this would in due course be extracted.

The great variation between these estimates is perhaps less important than it seems, for there is little doubt that the undiscovered oil will be more costly than that now being exploited. But there is unfortunately no reliable way of predicting just how much more expensive it will be. The practical issue is simply that there are enough likely oil provinces for the oil industry to exploit in the decade ahead. And in that period, as in the past, there will be acquired a more accurate picture of what will be possible in future years than any now available. Yet even allowing for the dominance of the large oil fields, it is reasonable to suppose that the proved reserves in 1971, the equivalent of just over eight years of production, should be increased by between a third and a half simply to allow for the degree to which efficient recovery techniques have been made economic by the recent increase of price.

Whatever happens in the United States or elsewhere, however, for at least another decade the Middle East is bound to dominate the petroleum industry. Not merely is Middle East oil cheap to produce, but the reservoirs are large even by the standards of the richest fields in the United States. Thus the largest oil field in the United States, that in East Texas (discovered as recently as 1930), is estimated originally to have contained close on 750 million tons of oil, of which something like one-tenth remained in the ground at the end of 1973. The largest single oil field in North America, in Prudhoe Bay on the North Slope of Alaska, is thought conservatively to contain 1500 million tons of recoverable oil, and may yield twice as much, but even so will be only a tenth the size of the great reservoirs of the Middle East. Indeed, the 10 000 million tons of oil in the Burgan field of Kuwait is a substantial fraction of the total amount of oil so far extracted and consumed in the whole world and rather more than the most optimistic estimates of the amount of oil that will eventually be recovered from the North Sea.

The strictly commercial implications of the size of the oil reserves in the Middle East cannot be shaken off in the near future, say the next decade, however successful oil exploration may be elsewhere. That another oil-producing region comparable with the Middle East will be discovered is exceedingly unlikely – certainly it would be foolish to count on such good luck. But the increase of the price of OPEC oil effective from the beginning of 1974 means that oil companies operating elsewhere in the world will abruptly have found it profitable to carry through previously uneconomic developments. Moreover, it will be economic to develop oil reservoirs previously considered too small to be commercial, not merely in outlandish parts of the world but in well-trodden territory as well. Similarly, it will be profitable to increase the rate at which oil is extracted from known reservoirs – to drill more wells in Texas, for example. And finally, for strictly commercial reasons, sources of energy other than petroleum will be considered as serious alternatives to conventional petroleum.

The proved reserves of petroleum in the world as a whole are at present estimated to be 80 000 million tons, or twice the amount of crude oil so far used and enough for twenty years' production at the rate obtaining in 1973. Of this amount, 60 per cent is in the Middle East, and is enough to sustain production at the rate obtaining in 1973 for thirty years or more. But the oil remaining to be discovered is estimated even by conservative geologists to amount to 150 000 million tons. Much of it will, however, be expensive oil.

For technical reasons, natural gas, the other principal petroleum hydrocarbon, occupies a less sensitive place in the world's energy economy. The chief natural gas reservoirs are comparable in energy content with the principal oil reservoirs of the world. The quality of the gas as a source of energy can vary enormously, both in the proportions of inert gases (such as carbon dioxide and nitrogen) which they contain, and in their content of hydrogen sulphide. The chief hydrocarbon is methane, chemically the simplest of them all, and the most difficult to liquefy, but there are usually substantial proportions of other hydrocarbon gases such as ethane, butane and the like. Most natural gas reservoirs yield not merely gas but substantial amounts of liquid hydrocarbons – in the United States, natural gas wells yield close on 100 million tons a year of 'natural gas liquids', which are both a valuable supplement to crude oil production and an especially valuable source of chemical feedstocks. Natural gas is also produced as a by-product at most oil wells. The Burgan field in Kuwait, which produced more than 150 million tons of oil in 1971, also produced natural gas energetically equivalent to 13 million tons of oil, roughly half of which was disposed of by the well-head flares which serve to pick out Kuwait on photographs of the Earth orbiting satellites.

Much less is known of the economic potential of natural gas than of liquid petroleum. For one thing, outside the United States and Canada, exploration has until recently been desultory. There are also technical uncertainties, ranging from that of knowing whether it will ever be feasible to use nuclear explosions for releasing natural gas from tight formations like those beneath the western states in North America, to those of building and operating vessels to transport liquefied natural gas from oil-producing countries. Even so, it seems to be accepted in the United States, where deep offshore production of natural gas has only just begun (with the first well in the Gulf of Mexico in 1974) that the gas so far discovered accounts for only 30 per cent of what will eventually be found. In other words, the exploitable natural gas that remains to be discovered in the United States accounts for the equivalent of 40 000 million tons of petroleum (1000 cubic metres of natural gas=0.9 tons of crude oil in energy potential). On paper at least, this is enough to last a century at the present rate of production. Outside the United States, estimates of natural gas remaining in the ground are more uncertain, but it would be surprising if American expectations were entirely inapplicable. As always, what will matter is not how much gas there is, but how much it costs.

In a world in which petroleum, oil and natural gas account for more than a third of energy consumption, and since they are, in geological terms, intrinsically scarce, it is inevitable that they should be at the centre of current speculations about energy supplies. But other fossil fuels are much more abundant. Several deposits of heavy oil, too viscous to be extracted with present techniques, are known to exist in the Western Hemisphere, principally in Canada, the United States and Venezuela. The United States is known to contain some 25 000 million tons of these materials, which are also known to exist in some unusual places in the Middle East – Turkey, for example – and there is no reason why these should not in due course be worked. Inevitably, however, deposits like these can contribute only marginally to the energy economy of future decades – the quantities involved are small, recovery is likely always to be incomplete and costs will be high.

The tar sands, chiefly found in North America, are a more hopeful supplement to existing oil resources. The most spectacular tar-sand deposit is that of the Athabasca River valley in Canada, where some 13 000 square miles of territory is underlain by sandstone 150 feet thick containing as much as 10 per cent or more by weight of viscous petroleum hydrocarbons. Something like 100 000 million tons of oil is contained in all but the leanest (less than 2 per cent) of this sandstone, and it has been estimated that more than 40 per cent of this may eventually be recovered, which is equivalent to roughly half the proved reserves of conventional oil, or to something like fifteen years of the world's present consumption of conventional petroleum.

Not even the riches of Athabasca, however, promise much more than an amelioration of the difficulties that have arisen since 1970. First, the amounts of oil in the tar sands, large though they may be, are not sufficient in themselves to transform the balance between the different fossil fuels. Secondly, the technology of oil extraction is complicated and still unproven. The first plant built 250 miles north of Edmonton has been operating since 1967 and uses a simple process in which oil is extracted from the sandstone by hot water. The oil has to be hydrogenated to produce a liquid comparable in its physical properties with natural petroleum. Capital costs are considerable, but it is reckoned that oil can nevertheless be produced for about $7.00 a barrel. In other words, Athabasca oil should now be competitive in price, which explains why the oil companies were at the beginning of 1974 eager to lease tracts of the Athabasca tar sands from the provincial government of Alberta. Since then, their enthusiasm has waned, partly because of the effects of high interest rates on the cost of capital developments, partly because the government of Alberta and the federal government of Canada have turned out to be as grasping as any OPEC state in what they require of the oil companies. The result is that production is unlikely to grow quickly, but the output of Athabasca may amount to 50 million tons of oil by 1985 and 250 million tons a year by the end of the century, though even then the output of this remarkable geological structure will be less than half the current production of conventional petroleum in the United States.

The oil shales of the world are a different matter. In principle, the quantities of hydrocarbons that might be recovered from them, or even from those of Utah, Wyoming and Colorado alone, are large enough to change the balance between the conventional and the unconventional petroleum hydrocarbons. There is therefore a touch of irony in recalling that the petroleum industry in Europe began in the eighteenth century with several successful attempts to win a liquid fuel from oil shale.

Oil shales have, of course, been formed geologically by the incorporation of organic matter into sedimentary rocks. The great deposits in the Rocky Mountain states accumulated in a large lake system in Eocene times, 40 to 50 million years ago. The sandstone is in some places as much as 1000 feet thick. The organic material itself is deficient in hydrogen compared with the conventional petroleum hydrocarbons, with the result that petroleum fluids can be obtained only by some form of thermal treatment. The four leases on tracts of land in the Rocky Mountains let to private developers at the beginning of 1974, two each in Colorado and Wyoming are reckoned to contain enough organic material to yield thirty US gallons of oil for each ton of rock. On the conservative basis that only deposits of oil shale at least fifteen feet thick and containing fifteen gallons of shale oil in each ton of rock are counted, the US Geological Survey has estimated that the Green River Formation spanning Colorado, Utah and Wyoming contains some 300 000 million tons of oil, or more than three times the total amount of conventional petroleum ever likely to be recovered from the world's oil wells; and the Green River Formation in the Rocky Mountains is only one of several potential sources of shale oil even in the United States. In the eastern and central United States, some 250 000 square miles of territory is underlain with a black marine shale known by a variety of names, but chiefly the 'Chattanooga Shale', which ranges in thickness up to 100 feet, and in oil content up to fifteen US gallons per ton of rock. It has been estimated that the total amount of oil in oil shales throughout the world may amount to 50 million million tons.

These quantities are so large that their size is irrelevant to the pattern of the world's energy consumption, now or in future centuries. As always, what matters is not physical availability but the cost at which liquid fuels can be produced as the end product. By the early 1980s, there should be enough experience of how the oil shale plants to be built in the Rockies have functioned, but the US Department of the

Interior is, for the time being, committed to license only enough plants on publicly owned land to allow the production of 125 million tons of oil a year in 1985. Only after that will the oil shale industry be in a position to produce amounts of energy comparable with the likely increase of energy consumption in the United States in the period 1975–85. The initial cost of shale oil was estimated at the end of 1973 at between $35 and $40 per ton of oil substitute or synthetic crude oil, sometimes called 'syncrude', but operating experience is expected to reduce this cost by 5 or 10 per cent. Petroleum companies interested in oil shale leases tend to higher estimates of cost, upwards of $50 per ton of product. At least four versions of retorts have been developed to the pilot-plant stage, but there is no immediate prospect of commercial success for techniques for extracting petroleum from oil shales by vaporizing organic material by underground heat treatment of the oil shales while they are still buried beneath their overburden.

The development of the Rocky Mountain oil shales is likely further to be decelerated by environmental restraints, chiefly the need to dispose of very large volumes of waste material, the need for large quantities of water in a semi-arid region of the United States, and the inevitable changes that oil shale development will bring to a previously unspoiled part of the country. Capital costs as such are likely to be less a restraint in the immediate future than the difficulty of constructing retorting plants in large numbers, but the National Petroleum Council has estimated that capital costs will range from $5000 to $7500 per barrel per day of oil produced, or between two and three times as much as the average cost of developing conventional oil wells in the United States. Allowing for a 15 per cent rate of return on capital, the council calculates that the bare cost of producing shale oil will exceed $5 per barrel, which is not essentially different from other estimates of cost.

With both these unconventional sources of petroleum products, North America is again the only region in which even guesses of future potential can be made, and the reasons are at present more technical than economic. For the time being, it would be wrong to regard the development of the

Rocky Mountain oil shales and the Athabasca tar sands as more than substantial programmes of technical development. Elsewhere in the world, exploration for oil shale and tar sand deposits has hardly begun, even though it seems that the Chinese People's Republic produces something like the equivalent of 40 million tons a year of oil from the oil shales of Manchuria. In the long run, it will probably turn out that the economic value of materials like this will depend not only on the development of economical techniques for extracting the petroleum but on the small but significant amounts of potentially valuable minerals they contain. The Chattanooga shales, for example, contain significant proportions of uranium.

The increased price of petroleum since the beginning of 1974 has also throughout the world renewed interest in the coal industry. In the United States, this happens to be entirely sensible, but there is a danger that other countries will be stimulated to follow suit in circumstances which are entirely inappropriate. The truth is that the United States is exceedingly well-endowed with rich coal deposits which can be mined economically. Jevons, a century ago, was right. But again, what matters about the coal reserves of the United States is not the amount of coal beneath the ground but the ease and cheapness with which it can be extracted. American coal reserves, identified and still to be discovered, are reckoned to amount to 3.2 million million tons, about half of which could probably be recovered with known technology. In terms of energy content, this is roughly twice the amount of oil likely to be discovered in the whole world, and four times the amount of oil likely to be recoverable from conventional oil wells.

One half of the United States reserves is found in coal seams covered by less than 1000 feet of overburden, and one quarter of this is in thick seams, defined by the US Bureau of Mines as thicker than 3.5 feet for high-quality coal and thicker than 10 feet for poorer coal. Some of the richest coal fields not yet exploited are found in the northern Great Plains region, chiefly the states of Wyoming, Nebraska, Montana and the Dakotas, a region hitherto so remote from the principal markets for energy that the coal has not previously been exploited. Other important deposits lie further south, in New Mexico and Nevada. Mining costs for this material are estimated at no more than $2.00 per ton of coal, and the transport of energy to the centres at which it will be consumed will probably be accomplished by conversion to electricity, synthetic gas and even liquid oil. In present circumstances, the impediments to the development of these resources stem from environmental considerations, chiefly the development of a bulk mining industry in previously unspoiled parts of the United States.

Not all regions in the world are so fortunate. In Western Europe, where low-grade oil shale is plentiful, rich oil shale virtually non-existent and tar sands unknown, the coal seams are thin and deep down compared with those of the United States. In Europe, the productivity of miners working in deep mines averages only a little more than two tons per day, compared with fifteen tons from the bituminous mines of the United States. Large reserves of coal remain in the ground, but their size is not an accurate guide to their usefulness. There remains a possibility that some radical new technique for extracting the energy of coal without mining it, among which underground gasification is the favourite, might make accessible a much greater proportion of the coal reserves of Europe, but the prospects are not bright.

In the Soviet Union and China, on the other hand, coal reserves are actually greater than those of the United States and are potentially as productive. Asia as a whole is thought to include two-thirds of the world's coal resources, recoverable and otherwise. There are geological reasons why other continents – Africa, for example – should contain much smaller (if still substantial) quantities of coal. In the world as a whole, it has been estimated that total coal resources were originally close on 17 million million tons, of which 9.5 million million tons had been identified in 1972. This estimate includes only coal in seams thicker than one foot and at a depth of less than 6000 feet, and these are very large amounts, enough for 300 years of consumption at the present rate. As with the other fossil fuels, however, the quantities of coal in the surface of the earth are as irrelevant to the immediate or even the long-term development of the world's fuel economy as is the limited size of the known reserves of petroleum hydrocarbons to the prospect that there will be physical shortage of energy in the near future. All that matters is what it costs, and geological uncertainties are here unfortunately complicated by uncertainties about the future cost of labour.

But is not the earth finite? Will there not come a time when all deposits of petroleum, coal and other hydrocarbon fuels will have to be worked for their energy content? And will it not then be the case that the cost of energy, essentially determined by the labour cost of extracting fossil fuels from deposits now regarded as hopelessly uneconomic, will be much greater than it is at present, with serious implications for the cost of the products of manufacturing industry and for the use that society makes of energy in all forms? These questions imply that the future of society would be compromised by the finiteness of the earth's crust. In reality, there are alternative sources of energy which are physically almost literally inexhaustible and large and uniform enough to suggest that production costs will be stable.

Nuclear energy is one. Nuclear power stations of the kind now in service use uranium as a fuel, so that on the face of things the amount of energy eventually recoverable from nuclear power stations would seem to be limited by the quantity of uranium in the earth's crust, itself in principle a finite amount. One defect in this simple calculation, however, is that the amount of energy recovered from, say, a ton of uranium depends critically on the kind of nuclear reactor in which it is used as fuel. For the nuclear reactors now in service, the amount of energy (invariably as electricity) obtained from a ton of uranium is the equivalent of what could be obtained by burning in a conventional electricity power station between 6000 and 60 000 tons of petroleum, according to the type of reactor. When fast reactors are in service, the upper limit will be increased to roughly three million tons of petroleum.

Two conclusions follow. First, the development of nuclear reactors and the use of uranium as a source of energy has at one stroke increased the accessible sources of energy in the earth's crust by an amount comparable with the coal deposits that have sustained industrial development and growth for the past two centuries. The quantities of uranium in relatively rich deposits already found and which can be mined at modest cost are reckoned to be 1.5 million tons or thereabouts, in principle perhaps the equivalent of 2 million million tons of petroleum. But high-cost uranium, even that dissolved in sea water, could be used if necessary, while thorium is an alternative source of energy to uranium with roughly the same abundance. Even without the prospect of thermonuclear fusion at some point in the next century, foreseeable developments in nuclear power have brought about a massive increase in the energy resources of the earth's crust.

Secondly, the economics of nuclear power are such that the cost of the electricity produced is much more closely determined by the capital cost of building reactors than by the market price of uranium, at least at present uranium prices. The consequence is that there is a sense in which the source of the electricity produced by nuclear power stations is not the ostensible fuel – uranium, plutonium, thorium or whatever – but the technology and the capital investment that go to the construction of nuclear power stations. In the 1950s, these characteristics were the source of many confident predictions that nuclear power would be abundant and cheap. Fears of physical exhaustion are indeed even less appropriate than with the other fossil fuels, but events have also shown that nuclear power is not necessarily cheap – electricity from nuclear power in the industrialized nations of the West is cheaper than electricity from oil at a tax-paid cost of $7.00 a barrel, but capital costs are high and, in the nature of things, are payments in advance for the benefit of energy consumers up to a quarter of a century hence.

Precisely what will be the course of development of nuclear power will, in the decades ahead, depend not so much on the availability of nuclear fuels as on financial considerations, chiefly the availability of capital and the extent to which the need for this will be modified by technical innovation. As things stand, the cost of building nuclear power stations works out at roughly $15 000 for the energetic equivalent of one barrel of oil a day, or more than a hundred times the development cost of cheap oil from the Middle East. Safety is another serious though not insuperable problem. There are several grounds for caution. All nuclear plants release measurable amounts of radioactivity into the environment. The safe storage of radioactive waste materials from reprocessing plants is another problem with formidable characteristics – safety must be assured for centuries on end. Finally, there are problems which stem from the possibility that plutonium or other nuclear explosives might be put to clandestine use as nuclear weapons. All of these problems deserve (and have, for that matter, received) careful attention. Each of them requires the development of new or more stringent administrative procedures and international agreements. None of them is, however, insoluble.

But why not have the best of both worlds? Why not rely instead on new sources of energy free from hazards – solar energy, geothermal energy and even thermonuclear energy? The short answer is that it is not now or in the foreseeable future (say the next two decades) feasible to meet the growing demand for energy by these means, and there must be serious doubts whether the capital cost of these alternative sources of energy will ever be lower than, for example, those of nuclear power. The contribution they will make to future energy supplies is certain to be negligible for several decades, and may always be small.

So how will the importance of different sources of energy change in the years ahead? It is natural for human beings to want to know what the future will be like – there is now even a temptation to believe that it is possible to elect governments that will ensure that some predetermined future will come about. But one of the consequences of the multiplication of the technical means of producing energy in the past two decades is that it has become economically hazardous to anticipate how quite small changes in people's expectations may have a profound influence on the most economical pattern of energy consumption. What this implies is that an energy policy should not be a device for saying what proportions of oil and coal should be used at some point in the future, but a device for allowing the most economic balance between different fuels to be found.

4
Brittle market

The rapid increase of the price of OPEC oil between 1970 and 1974, and the disturbance which this has caused among the oil-consuming states, has its origin in the ways in which the oil-consuming states have managed their supplies of energy in the past quarter of a century and before. Europe and Japan were brought face to face with the prospect of an acute shortage of energy in the early 1950s, when it became apparent that indigenous supplies of coal would be insufficient for current needs, let alone for the industrial expansion to which they looked forward. In Europe, the coal mines were under-capitalized, but it was also clear that the cost of coal would be continually increased by the increasingly dominant cost of labour in its extraction. For what it is worth, the coal shortage of the winter of 1946–7 was more serious in its economic effects than anything that has happened since in Britain.

Except in the United States, where indigenous supplies of oil seemed adequate for all future needs, the result was that the governments of industrialized countries were persuaded by their experience in the years immediately after the Second World War that it was their responsibility to forecast the demand for energy in the years ahead, and then somehow to satisfy themselves that the demand could be met by foreseeable supplies of fuel. In many countries, in Britain, for example, this exercise has been mistaken for a policy on energy. And now, in the wake of the oil-price increase, the fashion has spread to the United States.

The truth is that forecasting the future has a valuable part to play in the management of a fuel economy, but chiefly as a means of telling how current assumptions are likely to be falsified by events. As was equally true a quarter of a century ago, forecasts will frequently suggest that the time has come for strenuous efforts to be made to develop alternative sources of energy. But the essential ingredients of a nation's energy policy must be the steps taken to make sure that different kinds of energy are sold at prices which reflect the cost of production from the most expensive sources of energy – the prices which ensure that the most economical use is made of such resources as there are. As will be seen, most governments have shrunk from this part of their responsibility, and have indeed taken deliberate steps to distort the proper balance between alternative fuels.

The pitfalls of forecasting include some conspicious ways in which the economic balance of market forces can be deliberately upset. Forecasting of some kind must play a part in any attempt to formulate a policy for energy, and so it is unfortunate that forecasters should be figures of fun even more often than economists in general. In reality, it is neither surprising nor necessarily scandalous that different projections into the future should often differ or be falsified by events. Forecasts differ when they are based on different assumptions or make use of data from widely different sources. They are invariably falsified when circumstances change to make those assumptions no longer valid. Indeed, a forecast of a nation's energy consumption for some decades ahead should, more properly, be regarded as a way of deciding how the circumstances implied in the assumptions ought to be modified by government policies rather than as a vision of what the future will really be like. Regrettably, forecasters too often invite ridicule by claiming more for their projections than is reasonable.

If forecasts are to help with the formation of sensible energy policies, it is essential that they should forecast both the demand for energy and the availability of its supply. Because energy consumption is related to the general level of economic activity in a country, forecasts of the future need for energy invariably entail, explicitly or otherwise, assumptions about the rate of economic growth, measured by the GNP; or, more sensibly some would say, by the Gross Domestic Product

(GDP). Because the link between the growth of energy consumption and economic growth can vary from one country to another, one nation's forecast is necessarily different from another's. And because the link between economic activity and energy consumption within a single nation can vary from one decade to another as a result of slow changes in the pattern of its economic activity, forecasts that assume nothing but a relationship between economic activity and energy consumption cannot be expected to be accurate over periods spanning several decades.

In principle, more sensitive forecasts of the growth of energy consumption can be arrived at by dealing separately with different sectors of an economy: manufacturing industry of different kinds, transport and domestic consumption, and the like. These forecasts have the virtue that they allow for the likelihood that some parts of the market for energy will, in due course, be saturated (a second domestic freezer in every home is unlikely), but they also suffer from the defect that they cannot predict new uses for energy of the kinds continually being created (the domestic freezer was a rarity when the President's Commission on Natural Resources, the Paley Commission, produced its forecast of future energy needs in the United States in 1952).

In the past few years, and especially since it has become possible not merely to build but also to operate electronic computers, more sophisticated ways of tackling these problems of forecasting have become possible. What is called 'systems analysis' has arrived. The principle is simple enough. It is now possible so to programme a computer that its behaviour will simulate that of some real system – a national economy, the functioning of an electricity supply system or even the behaviour of an international market in energy. If the computer is then provided with data that describe the demands being made on the real system, it will disgorge a description of how it will respond.

The difficulty, as with all computer programmes, is that the simulated system is no better than the assumptions on which it is constructed. So far, models of national economies have been valuable chiefly for the pointers they provide to the research that must be carried out before a really accurate simulation can be constructed on a still larger computer. Much the same is true of the more rudimentary computer models of the world's energy economy now being constructed in government departments and academic institutions. On paper and in theory, it should be possible to calculate how the demand for energy in different forms will develop over the years or decades, given assumptions about the pace and character of economic growth and the availability of different kinds of energy at various prices. In practice, if only for the time being, the task is formidable.

The technique of systems analysis has been shown to be a powerful tool in deciding how to operate an electricity supply network from day to day, or how an oil company should divide its production between one oil field and another. On the other hand, the technique's application to the task of predicting how the balance between the world's need for and supply of energy will be balanced in the decades ahead is so limited by ignorance of basic information (including the likely effect of increased prices on the consumption of gasoline in the United States), by procedural inadequacies (including the inability of most existing models to take account of the stepwise nature of development) and by the empirical truth that economic development is accident-prone, that these modern techniques can, for some time to come, be intrinsically no better than more old-fashioned methods of forecasting.

In spite of such difficulties, forecasts of future energy consumption for individual countries, for regions and for the world as a whole are now much in vogue. They are not to be scorned, for they do at least have the virtue of suggesting something of the magnitude of the likely demand for energy in the decades ahead, provided people's expectations of the growth of economic prosperity are sustained and that the structural changes in the economy which are certain to come about are not so great as to produce substantial changes in the pattern of energy consumption.

One kind of forecast of future needs is, however, deeply suspect. In the past few years, it has become fashionable to calculate the future need for energy (or, for that matter, the future size of the human population) by assuming that the numerical rate of growth in some past interval of time will be repeated indefinitely into the future. For the world as a whole, for example, energy consumption in the 1960s was growing at a rate which corresponded to a doubling of energy consumption every sixteen years. So is it not prudent to suppose that, in 1990, energy consumption will be twice what it was in 1974, that it will double again in the sixteen succeeding years and that, by the end of the twenty-first century, the world's consumption of energy will be roughly 250 times as great as it is now?

The fallacies in these arguments, often dignified by the name 'exponential growth', are several. They take no account, for example, of the likelihood that the developing countries in which energy consumption is at present growing most rapidly in percentage terms will, in due course, acquire a measure of economic maturity, and hence the looser connection between energy consumption and economic growth that characterizes the advanced economies of the world. Nor does it take account of the possibility, even likelihood, that the pace of economic growth throughout the 1960s was exceptional. The overwhelming error of forecasts based on such primitive projections is, however, that they make no allowance for the possibility that the demand for energy will be at least to some extent modified by the technical difficulties of providing it, and by the cost at which it is made available. The facile doctrine of exponential growth leads inexorably to absurdity, for at some stage along the line of development, demand must always outstrip supply.

Forecasting the supply of energy is a more difficult and much neglected activity. Indeed, those who forecast the demand for energy are frequently content to assume no more than that supply, in the old-fashioned economic sense, will somehow automatically increase to match demand. Given the importance of energy to the industrialized economies, this might seem a reasonable position. If energy is as indispensable to economic activity as history suggests, surely entrepreneurs will come forward to meet the needs of those who use energy. And in any case, the relationship between supply and demand will in part at least be balanced by increases of price when supplies are short.

This is what should happen in an ideal world, but the real

world of energy policy is far from ideal. For one thing, the energy supply industries are almost always industrial activities whose output cannot be quickly increased. However much the price of electricity may be increased, the output of an electricity generating plant cannot be greater than the capacity of its generating sets. The output of a deep coal mine is in many practical ways limited by the capacity of its facilities for hauling coal away from the coal face. It follows that plans to meet the future need for energy must entail plans to build new production facilities which will almost always take at least three years (even for the development of a new open-cast mining operation) and ten years or more for a novel type of nuclear plant or a large hydroelectric scheme. The supply of energy, in other words, is less flexible than the supply of Coca-Cola. But there are other more serious reasons why supply and demand do not automatically match each other, not the least of which is that governments have traditionally considered it to be their job to regulate the price of energy, a frequently legitimate responsibility too often interpreted as a duty to keep prices below their economic level.

With all these reservations, the forecasts of even the recent past are interesting in themselves as well as pointers to the way energy policies have failed to adapt themselves to urgent needs. Two decades ago, in the summer of 1955, the first United Nations Conference on the Peaceful Uses of Atomic Energy provided a striking warning of our present troubles. The argument, most clearly expressed by two British economists, E. A. G. Robinson and G. H. Daniel, was that the world's energy consumption would continue to increase at a rate in some sense commensurate with the increase of prosperity. In a paper entitled 'The World's Need for a New Source of Energy', Robinson and Daniel set out modestly to estimate the future growth of the demand for energy. Historically, consumption had grown at 2 per cent a year compound since 1929, but various factors – the rate of growth of the world's population as well as the industrialization of the developing countries – gave weight to the belief that the future rate of growth would be somewhat faster, certainly above 2 per cent a year, possibly as much as 3 per cent a year. The forecast, for what it is worth, was too modest; between 1950 and 1970, the world's consumption of energy in all forms grew at an average annual rate of 4.5 per cent.

The essence of the argument by which Robinson and Daniel reached their conclusion was that the world's reserves of conventional fossil fuels, known and in prospect, were unlikely to be sufficient to meet even their modest estimates of how demand would grow. Their scepticism about the difficulty of turning underground reserves of coal into fuel has been justified by the depressing experience of the coal industries of Western Europe in the past two decades. Although their estimates of the ultimate reserves of petroleum were only two-thirds of the estimates now current (see Chapter 3), their estimate of the consumption of energy in 1970 (the equivalent of between 2530 and 3460 million tons of oil) was substantially less than the 4916 million tons of oil or its equivalent actually consumed in that year.

> It will, I think, be apparent that, even if we make conservative estimates of economic growth and of the future increases of demand for energy, the world is not far distant, measured in the units of time in which we think of the history of nations and even of the lives of individuals, from the moment when, in the absence of a new source, scarcity of fuel will begin to create serious problems. The remarkable material progress of the human race during the past two centuries has largely sprung from the opportunities presented first by coal and steam power and more recently by oil and hydroelectricity to supplement human and animal muscles with other forms of energy. The wealth of the wealthier nations derives from the fact that they employ many times as much horse power per worker. In the United States the supply per head of energy from fuel and water power is some ten times that of some of the poorer countries of Europe, and a hundred times that of some of the poorest countries of the East. Economic development is largely a matter of adding to the horse power available to assist the people who at present are poor because of lack of it. There is no task more important than to ensure that the dramatic advance of the human race over the past two hundred years is not reversed over the next two hundred years because of the exhaustion of the fuels which gave the opportunity for it.

The first International Conference on the Peaceful Uses of

Atomic Energy was a sounding board for anxieties about the future supply of energy that had already begun to surface. In Britain, the coal shortages of the post-war years stimulated a spate of introspection about the future of the coal industry that culminated, in 1955, in an open recognition by government that only the rapid development of nuclear power would allow a future sustained development of the British economy. The same point was made in 1956 by a committee appointed by the Organization for European Economic Cooperation (OEEC), which has become the Organization for European Cooperation and Development (OECD) with the accession of Canada, Japan, the United States and Australia. The committee also argued that a policy of dependence on petroleum supplies from the Middle East would be strategically unwise. In the United States, the case for nuclear power was not nearly as strong, although the Paley Commission in 1952 had foreseen that the increasing trend of prices rather than physical shortages of coal and petroleum would make the United States dependent on imported fuel. Nevertheless, the United States, like other industrialized nations, embarked on an ambitious programme for the development of nuclear power. To all appearances, in other words, there was by the mid-1950s an open recognition that then existing sources of energy would be inadequate, that governments in industrialized countries must develop nuclear power as an alternative, and that there was no time to waste.

Why then did the nations of Western Europe come to rely on oil from the Middle East instead? There is, of course, no simple answer, but some partial explanations may be picked out. First, the prophets of nuclear energy had overplayed their hand – nuclear reactors turned out to be more difficult and more expensive to build than had been thought, for reasons to be dealt with later. Secondly, the decline of the prices of some other sources of energy, which had persisted for the best part of half a century, continued, especially in the manufacture of electricity from conventional fuels, no doubt in part because of the threat of competition from nuclear energy. And although pessimism about the feasibility of extracting coal from underground reserves was frequently justified by events, with the result that the infant European Iron and Steel Community embarked on a policy whose consequence was the deliberate contraction of the coal industry of Western Europe (Britain excluded), the reserves of oil in the Middle East, and the ease with which they could be exploited, persuaded Europe and Japan that cheap oil from the Middle East was the best source of energy for the 1960s, a view reinforced by the discovery of natural gas in the Netherlands (at Groningen) and France (at Laq).

In 1960, a committee appointed by the OEEC argued in a report entitled *Towards a New Energy Pattern in Europe* that the rapid development of the oil fields of the Middle East had created a new situation in which the best energy policy was a policy of playing the market, buying energy from wherever it could be had more cheaply. The committee said:

> There is unlikely to be any persistent long-term shortage of supplies of primary energy by 1975. While the need to create new sources of energy [nuclear power] is a real one for a much longer period than we have under review, it does not seem likely that shortages of oil or other supplies will make themselves felt in an acute form by 1975. Member countries should not therefore frame their energy policies on the assumption of a probable 'gap' in European energy supplies, in the sense that insufficient energy would be available at the source.

The committee gave its opinion that the security of the supply of oil could be ensured by commercial agreements without unduly increasing the price of imported oil. Member governments were urged to 'review their plans for the development of nuclear energy in the light of the changed long-term position regarding the likelihood of an energy gap and the probable trends of technical and economic development of this form of energy'.

The errors running through this once influential policy recommendation are significant, not merely as explanations of what has since gone wrong, but also because the shortsightedness that they embody persists, and is in danger of undermining the policies now being formulated for the years ahead.

First, however, it must be acknowledged that it was entirely sensible of the OEEC committee to argue that countries are mistaken if they embark on policies whose chief objective is self-sufficiency in the supply of energy. Invariably, the result is to increase costs unnecessarily. In 1960, Europe was using the equivalent of 600 million tons of oil a year, so that a saving of even as little as $5.00 a ton by substituting Middle East oil for coal could bring important advantages. This was in any case a time when, as will be seen, oil prices were falling slowly but significantly while the technological problems of bulk transport in large tankers were well on the way to solution. The advantages of a policy of dependence on Middle East oil were, for Europe, further strengthened by the way in which a large part of the capital cost of development in the Middle East was financed from the resources of American oil companies.

The errors of the policy advocated by the OEEC, and in practice adopted by most countries in Western Europe as well as by Japan, are subtle, but not easily forgivable. Like most other arguments about energy policy in the late 1950s, the OEEC report was innocent of considerations of the likely movement of the price of oil. To be sure, the period was full of gloomy prophecies about the increasing trend of the cost of mining coal, but nobody appears to have considered that a transformation of the energy economy of Europe from a condition in which coal was the staple fuel and oil a relatively unimportant supplement to one in which oil was king would ultimately have a drastic consequence for prices.

The point is simple enough. If there are two sources for the supply of a single commodity (in this case energy), and if everybody is behaving sensibly, either the more expensive will be dispensed with or, if supply is insufficient, both will be used and the price of the low-cost commodity will rise to meet the price at which the more expensive is being sold. A little reflection will show that this simple principle is not merely an immense benefit for the man who happens to be selling the low-cost commodity but also for his competitor, who is at least able to keep on fighting for his share of the market and who can hope to stay in business long enough to recover his capital investment.

In 1960, in Europe, the high-cost fuel was coal, the low-cost fuel was oil. It is true that both fuels had special uses – coal of certain kinds in making steel, petroleum as a source of motor spirit – in which they could not be substituted for each other, while for each of them there was a spread of production costs and also of quality. The moral for the OEEC committee and for those who listened to it should, however, have been that unless they were seriously contemplating a situation in which coal would cease to be used as a general source of energy, as for generating electricity, together with the waste of past capital investments that would naturally follow, they should have allowed the price of oil to increase so as to match that of coal. If oil companies had sought to increase their share of the market by manipulating their prices downwards, they could at least have been required to back their implicit promises with long-term fixed price contracts, especially for the electricity utilities that built power stations to burn oil exclusively. Ultimately, taxpayers would expect to get the benefits of cheap oil not in their electricity bills but from the corporation taxes paid by the oil companies. And in the last resort, it would always have been possible for governments to ensure some kind of stability by the clumsy device of excise taxes.

How did it come about that the governments of Western Europe failed to follow the precepts of the elementary economic textbooks? The simple answer is that they did not look far enough ahead. For practical purposes, they assumed that the oil companies would be able to keep their implicit promise and provide a supply of oil at roughly constant prices for newly built industrial plant. They took no steps to follow the injunction of the OEEC report that 'safeguards from possible future interruptions should be sought' and, indeed, had no means of embarking on such a task. As events have shown, the oil companies themselves were in no position to keep their implicit promise, and one reading of the rapid increase of oil prices in the past few years is that OPEC countries have now done what the elementary laws of economics would have accomplished anyway.

The other serious error in the policies that evolved in Europe in the 1960s was the failure to appreciate that the prospective 'gap' between consumption and supply which had startled governments only five years earlier could not be made to vanish indefinitely. Although the proved reserves of the

Middle East itself, already known to be large in 1955, were greatly extended in the following five years, the growth of reserves was not nearly great enough to suggest that 'the world's need of a new source of energy' had gone for good. Sooner or later, and sooner, as it turned out, nuclear power would be necessary. But even in the 1960s, European governments should have recognized that earlier over-optimism over the speed with which nuclear power might be developed (see Chapter 9) was an argument for even more vigorous research and development, not for indifference or positive neglect. It should also have been plain that even if the point at which nuclear power would be needed in bulk was hard to predict, the existence of a well-developed technology for an alternative source of energy to petroleum would have been a powerful counter in the argument about prices with OPEC (first formed in 1960) that was bound to come.

Although the thoughtless adoption of policies of cheap oil was a large part of the trouble, the errors of judgement that gave rise to them have recurred in succeeding years. The price of natural gas from the Dutch field at Groningen was set so far below the equivalent price of oil, and five-year and ten-year contracts were made so lightly with neighbouring countries, that the field is likely to be exhausted long before those who have been induced to consume the natural gas have had the full benefit of the capital they have invested in gas-burning appliances. Much the same is true of the development since the mid-1960s of the gas fields in the southern North Sea. The British Gas Corporation (as it now is) was so successful in persuading potential users to convert to gas, that by 1974 it was being forced to buy gas expensively (but just how expensively it would not say) from the Norwegian sector of the Frigg Field to make up for the tailing off of supplies from the southern North Sea likely to set in from 1977.

If the energy policies of Western Europe were clearly detached from reality by 1960, those of the United States had already for several years been suffused with an air of fantasy. Largely because the best-endowed country in the world has been, at least since the time of Theodore Roosevelt, the country most given to formal studies of the adequacy of its natural resources, the sorry record of error, confusion and contradiction is copiously documented. It is also richer and longer than any other.

Traditionally, American policies on mineral exploitation have been designed to stimulate production, even if one result is that materials are sold at prices higher than would strictly be necessary. Mineral rights on government lands have been sold at knock-down prices, exploration companies have been given grants and subsidies to help them venture into territory that would, on strictly commercial grounds, be unprofitable.

In the oil industry, the encouragement of exploration has for most of this century been accomplished by the device of depletion allowances, the system under which American oil companies are allowed by the US Government to deduct a certain percentage of their total sales from their taxable income before working out their tax bills. The theory is that an oil well is a wasting asset, so that its owner is entitled to special consideration. But the theory is unconvincing.

A company that invests in machinery recognizes that if it wishes to stay in business, it will have to replace the machinery sooner or later, either because it has been made obsolete or because it has worn out, and tax systems throughout the world, not merely in the United States, allow for this by letting manufacturers make provision for the depreciation of their capital investment. The practical result has been that American oil companies have been able to exploit profitably oil fields that would otherwise have been uneconomic, and the overdrilling of the most productive oil fields has been encouraged. At least until 1970, American consumers were paying more for oil products than was strictly necessary, but enjoyed the illusion of plenty as recompense.

The American record of mismanagement of oil and energy policy goes back to the beginning of the modern petroleum industry in the United States, which dates from 1859, the year in which Edwin L. Drake, working as a contractor for financial interests, struck oil at a depth of seventy-nine feet in eastern Pennsylvania. Drake's backers had prepared the ground carefully, commissioning a study of the potential of the new industry which led them to think that they would find not merely a cheaper source of lamp oil but a diversity of other products such as paraffin wax and lubricants. The technique of drilling boreholes had been advancing rapidly, chiefly in the development of underground salt deposits, but Drake and the early pioneers obtained little help from geological studies, and the geophysical prospecting techniques which are now an indispensable preliminary to the search for oil were not yet developed. The possibility that petrol or gasoline would become almost indispensable in surface transport was not considered, for the internal combustion engine had not been invented. Petroleum products were not used on any scale for heating until the beginning of this century.

The success of Drake's first well was quickly followed by a spouting of oil derricks throughout what is now known as the Appalachian oil field, the sedimentary basin running south from Ontario through Ohio and Pennsylvania into West Virginia. Not everybody made a fortune, but the lamp oil market was transformed wherever it could be reached. At the beginning, petroleum from the Appalachian oil fields was distributed in wooden casks, which were floated down the Ohio and other rivers for further distribution by the canal system in the eastern states and the railroads reaching further west. The last link in the chain of distribution was almost invariably a horse and cart. Yet, by the 1860s and 1870s, lamp paraffin was meeting competition in the cities from the gas works and the incandescent mantles increasingly used for lighting elegant homes.

To begin with, the infant oil industry resembled very closely the metal extraction industries of the United States. The first prospectors were the temperamental inheritors of the gold rush of 1848. Their operations, always a gamble in the absence of geological understanding, were financed by personal savings and modest bankers' loans. Even when oil gushed up from a hole that had been sunk, there was always a danger that it might at once catch fire. And in the early days, the market for petroleum was almost as volatile as the product itself.

But these early years also saw the foundation of the first of the multinational oil companies whose success in rationalizing the production of the Appalachian oil field must surely have been a sign that the petroleum business is, for economic reasons, almost inevitably big business. In 1863, Mr John D. Rockefeller bought a refinery at Cleveland, Ohio, and made this the centrepiece of an industrial trust whose chief function was to ensure that oil producers feeding crude oil to the refinery at Cleveland would regulate the level of their production so as to maintain the prices at which refined products could be sold. The Standard Oil Company was able to survive the financial panic of 1873 and, indeed, to take over a good many of the leases on the earliest oil fields in the east when their owners found it necessary to sell for cash. Although Standard Oil was often rebuffed in its attempts to extend its writ outside its original field of operations, by the end of the nineteenth century the company was virtually unchallenged master of the United States petroleum industry. It succeeded by its open practice of modern management techniques and its capacity to think big. But it was a monopoly in the simplest sense, which is a part of the reason why Senator Sherman, an Ohio senator, was the principal architect of the anti-trust legislation of the 1890s which eventually led Standard Oil to find a refuge in New Jersey, known as Esso and later (after 1971) as Exxon. By the turn of the century, Standard Oil had extended its search for oil to Mexico, Venezuela and the Caribbean. In spite of the enforced splitting off of Standard Oil of California and Ohio (Sohio), it remains the largest oil company in the world.

The ambivalence of the United States towards monopoly or, simply, big business is a constant source of interest and surprise to outsiders. It is also a part of the reason why in the international petroleum trade the tripartite relationship between the producing governments, the consuming governments and the multinational oil companies has been at best confused. At the end of the nineteenth century, the industrial trusts in railroads, sugar and steel as well as oil were respected, even revered. They were held up, in the United States, as models of good management and how ambitious young men could make great fortunes. The legislation against monopoly was, for the best part of three decades, ineffectual. Mr Rockefeller's lawyer, Samuel Dodd, said that it was as pointless to attempt to control the trusts as 'to stay the formation of the clouds, the falling of the rains or the flowing of the streams'. Standard Oil's case was upheld by the Supreme Court in 1907, but Mr Justice Harlan, in a dissenting opinion, delivered a perceptive explanation of why American feelings on trusts (which are illegal) and big businesses (which are not necessarily illegal) run deep when he said:

> All who recall the conditions of the country in 1890 will remember that there was everywhere among the people a deep feeling of unrest. The nation had been rid of human slavery . . . but the conviction was universal that the country was in real danger form another kind of slavery, namely the slavery that would result from the aggregation of capital in the hands of a few, controlling for their own advantage exclusively the entire business of the country, including the production and sale of the necessities of life.

The development of the petroleum industry in the United States is important, not merely because of the influence it has had elsewhere, but also because the regulatory apparatus created by successive United States administrations has had a direct bearing on negotiations in the past half century between the 'majors', as the largest oil companies like to call themselves, and the leading oil states. At the beginning of this century, the pace of development was quickened by two events: the introduction of the internal combustion engine, which created a huge demand for gasoline, and the discovery in 1909 of the Spindletop well in Texas, which promised domestic supplies of petroleum much greater than could be had from the Appalachian field. Spindletop stood in relationship to the other oil fields of the United States as the oil fields of the Persian Gulf do now to those known elsewhere in the world.

Between 1900 and 1920, United States production of petroleum increased nearly tenfold, from just over 9 million tons to 86 million tons a year (when United States production was 86.5 per cent of world production). In 1920, the petroleum industry was startled by a forecast by the US Geological Survey that recoverable reserves in the United States would turn out to be no more than 1450 million tons, enough for only seventeen years of production at the then current rate. The result was an energetic hunt for oil, not merely in Texas, but also in California, and, in 1923, when the postwar recession was biting, an excess of production capacity and a catastrophic fall of petroleum prices.

The response of the state of Texas, with the connivance of the central administration, was the creation of a system of prorationing from Texan oil wells by means of which oil-well operators were required on a monthly basis to extract only a stated proportion of their productive capacity of crude. This device, copied closely in states such as Louisiana, has over later decades had two important consequences for the pattern of the United States petroleum industry. First, it has meant that those with oil-well leases have been tempted to drill many more holes in the ground than would strictly be needed, simply to secure a larger share of the amount of oil that could be sold each month. Secondly, it has kept in being economically marginal oil wells, known as 'stripper' wells, producing only a few barrels a day of oil but able by their number to set domestic oil prices at the wellhead at a much higher level than would otherwise have been necessary. Since the mid-1920s, in other words, the influence of the largest oil producer in the world has been to maintain domestic prices in the United States and so, by extension, international prices, at needlessly high levels.

If the United States had been, in the 1920s or 1930s, an entirely isolated oil-producing and oil-consuming economy, the fairy-tale economics of its system of oil regulations might at least have been held to have provided a measure of stability. So much was most clearly apparent in 1930, during the collapse of oil prices brought about partly by the great depression, partly by new discoveries in Texas and California and partly by the lawyers' paradise created in east Texas where the holders of oil-well leases set about capturing the oil beneath their neighbours' plots by pumping hard and by drilling oil wells on the slant. During this period, and until the Second World War, the United States was for practical purposes a self-contained energy economy, with domestic supplies augmented by imported oil from Mexico, Venezuela, the Caribbean and Canada in quantities not large enough to disturb the cosy domestic arrangements for the domestic supply of energy. The fact that petroleum prices (in excess of \$2.00 a barrel at the Gulf of Mexico) in the late 1930s were higher than they need have been was not a serious burden on American industry nor a cause of friction with the outside world. Before the Second World War, the principal consequences of United States policies on oil were the waste of domestic economic resources and the laying of the foundations of the still more fanciful system of regulation and control that has grown up since 1945.

Throughout the postwar period, the United States petroleum market has been carefully, even intricately, protected for a variety of reasons, some of them strategic, some of them merely devices for ensuring the continued operation of essentially uneconomic oil producers. With the growing interest of American oil companies in the Middle East in the years immediately after the Second World War, it would have been entirely natural for expensive oil from domestic oil wells to have been replaced by cheaper oil from the Middle East. Indeed, between 1947 and 1955, the percentage of imported petroleum refined in the United States increased, partly because of the activities of the major oil companies (principally Esso and Gulf) in the early postwar period, later because of the activities of the independent companies without overseas oil fields of their own who purchased crude in the Middle East and elsewhere from the concessionary companies.

To the domestic petroleum industry, these tendencies were a potential threat. In 1949, Congress reflected the anxiety of the oil and other energy industries in the United States by undertaking studies of the consequencies of 'foreign oil imports' on the petroleum and coal industries as well as their possible effects on the strategic independence of the United States. In 1953, the Texas Railroad Commission took a hand in regulating the affairs of the industry by requiring the oil companies to declare in advance their plans for importing oil. By February 1955, a Cabinet committee had concluded that oil imports were a threat to national security and the work that the Texas Railroad Commission had pioneered in restricting imports had been taken over by the Office of Defense Mobilization.

To begin with, the office attempted to control imports by voluntary agreements with the companies, but the growth of the independent marketing organizations in the United States, by then habituated to buying oil from producing companies in the Middle East, meant that voluntary control was ineffectual. On 1 April 1959, the administration exercised the powers already voted by Congress and imposed mandatory controls on oil imports. Throughout the 1960s, the legislation was used partly to discriminate between potential sources of imported oil in the United States – Canadian oil was most preferred, oil from Venezuela and the Caribbean came next. The Middle East, made to seem especially unreliable by the Suez crisis of 1956 and the closure of the Suez Canal in 1957, came at the bottom of the list. In the late 1960s, imports from Libya were encouraged, chiefly because of a low sulphur content, but by 1970 the mandatory controls on imports and the rapidly growing demand for petroleum in the United States had meant that the prorationing quotas fixed by the state regulatory commissions had risen to 100 per cent of available capacity. For practical purposes, the United States had used up its spare capacity.

The precise influence of the import policy on the more recent upheaval in the petroleum market cannot, of course, be categorically defined. One view is that the import restrictions were a great benefit, not merely for the United States but also for other industrialized nations; that, without them, European and Japanese consumers would have paid more for petroleum in the 1960s, but the United States would have been even more dependent on Middle East producers than it was at the end of 1973, when OPEC prices were dramatically increased.

This argument is mistaken. The import quotas, like the ineffectual voluntary schemes for import regulation which preceded them, had the effect of increasing the price of oil in the United States to a level at which the least economic domestic producers could just stay in business, and of decreasing the price at which oil was sold to consumers elsewhere. These artificial devices, while eradicating the surplus capacity there would otherwise have been in the major oil fields of the United States, also stimulated the rapid growth of oil consumption elsewhere. The United States, as the largest consumer of oil, would have been in a much stronger position to counter the increasingly ambitious claims of the oil-producing states had it not itself maintained the price of oil at an artificially high level and had it not thrown away the bargaining strength that would have derived from unused surplus capacity. In several important ways, the negotiating position of the United States was further weakened by the detailed operation of the import regulations, which towards the end of the 1960s were relaxed to allow increasing imports of residual fuel oil from refineries overseas, chiefly for use in heating plants and electricity power stations; the consequence has been a shortage of domestic refining capacity which will not be made good until the 1980s.

Over the years, United States policy on petroleum has had the effect of increasing prices on the domestic market. Paradoxically, policies on natural gas have worked in the opposite direction, with the result that natural gas has been sold in the United States at prices which do not reflect its value as a fuel and which have over-stimulated consumption, discouraged exploration and distorted its pattern of use.

The origins of the haphazard system of control which has distorted the natural gas market go back to the 1930s and efforts then made to control the prices of commodities shipped across state boundaries. Natural gas was an obvious candidate for regulation, for interstate pipelines are the most convenient way of shipping this material from the gas wells in the West to the markets on the east and west coasts. By the beginning of the Second World War, the Federal Power Commission had been empowered to control the prices at which the pipeline owners paid gas-well operators for their product as well as the prices they could charge the ultimate consumers, mostly local or regional gas companies throughout the United States. In applying both kinds of control, the Federal Power Commission has held fast to the notion that operators should be allowed to make a fixed return on their capital investments as well as recovering their direct operating costs, with the result that the wholesale price of natural gas in the United States has only recently exceeded $0.25 for 1000 cubic feet, the equivalent of oil at $1.40 a barrel.

The consequences have been predictable and serious. First, the domestic market for natural gas in the United States has grown quickly. The domestic and commercial sectors of the American economy increased their use of natural gas six times between 1947 and 1969, and during the same period there was a four-fold increase in the consumption of natural gas by power stations – a trend encouraged by the exemption of agreements between gas-well operators and electricity utilities from price regulation provided both are in the same state. Although the commercial history of the United States is hardly littered with bankruptcies among pipeline operators and natural gas producers, there is good reason to suppose that even with the system of depletion allowances which the latter (like the oil companies) enjoyed, natural gas production was not sufficiently profitable to compete with other forms of industrial investment. Through the 1960s, the annual rate of drilling for gas and the number of completed gas wells both declined by about a third. Since 1967, more natural gas has been consumed each year than has been discovered. Whatever the theoretical arguments about the undiscovered resources of natural gas in the United States, by the end of the 1960s government policies had created a situation in which natural gas production would inevitably decline, but in which consumers were protected from the economic costs of obtaining this resource and, for what it is worth, natural gas producers were protected from the truth by subsidies in the guise of depletion allowances.

Towards the end of the 1960s, the problem of balancing supply and demand for energy in the United States was further complicated by the environmental movement. Clean air and clean water may seem to be essential to the well-being of prosperous societies, but the United States embarked on its great programme of cleaning up the environment with hardly a passing thought for the way in which this would upset the already hazardous balance between the supply and demand for energy. Environmental controls affected the energy balance in three principal ways. First, the movement to reduce the content of lead in gasoline, under way on a voluntary basis by 1970 but then given the force of law, makes internal combustion engines less efficient and also increases energy consumption in oil refineries. The regulation of the emission from road vehicle exhaust systems, introduced progressively from 1970, implies that new vehicles originally to have been introduced in 1976 (but mercifully postponed) would have made automobiles 15 per cent less efficient in their use of energy than vehicles of similar size and weight marketed a decade earlier.

The most serious, even bizarre, consequence of the Clean Air Act has, however, been the way in which it has required the fifty states of the union to draw up plans for the regulation of air quality (chiefly concentrating on sulphur dioxide emissions from industrial plants, including electricity generating stations). When the individual State Implementation Plans were drawn together by the Office of Emergency Preparedness in 1972, it became apparent that restrictions on the sulphur content of coal would mean that, by 1975, American industrial plants needing 592 million tons of coal would be able to burn only 268 million tons of what would be available in the United States. Similarly, restrictions on the sulphur content of oil suggested to the Office of Emergency Preparedness that, between 1972 and 1975, it would be necessary to increase imports of low-sulphur oil (from OPEC members such as Libya) by between 15 million and 150 million tons a year.

In the long run, it is true, the United States can hope to have the best of both worlds by making arrangements to remove sulphur dioxide produced by burning high-sulphur domestic fuels from exhaust gases before these are discharged into the atmosphere. The difficulty is that the techniques are not yet fully developed, that they add considerably to the cost of industrial operations and that their implementation will entail capital costs not easily met in inflationary circumstances.

The case of the Alaskan pipeline has been a more glaring assault on reason. Since 1968, it has been clear that the North Slope of Alaska is capable of providing some 150 million tons of oil a year, perhaps a quarter of domestic production in the United States, but the scheme for building a pipeline to carry the oil to the Pacific coast of Alaska was delayed roughly four years by fears that the construction of the pipeline would cause unacceptable damage in Alaska. That there should have been a careful examination of this argument is beyond dispute. What is disreputable is that the United States Administration signally failed to leaven the case for building the pipeline with an informed calculation of the economic costs of a decision that the pipeline should not be built and that, when the case against the pipeline had become by 1972 a legal wrangle in the courts, it neglected to use its influence with Congress to amend the law until the summer of 1973, when the impending increase of OPEC prices was assured.

Why has the United States been so obdurately indifferent to the arithmetic and economics of energy supply? The explanation is partly political and partly psychological. Domestically, successive United States administrations have sought to placate both the ultimate consumers of energy and those who supply them with appliances (the automobile manufacturers, for example). This is to be achieved by ensuring that energy in its various forms is sold as cheaply as possible, while oil producers are compensated by tax concessions of various kinds. The result has been to foster the illusion that the United States is a land of plenty, even during the period of the rapid decline of the ratio of reserves to annual production which set in from 1965. Thus have the past ten years seen the adoption of environmental programmes which, however desirable in themselves, can only widen the gap between desire and reality. The gasoline and gas shortages endemic since 1970 are thus not so much a sign that natural resources are limited as that, lacking the operation of market forces, it is impossible to balance supply and demand without arbitrary restrictions imposed either by retailers or government.

The policies that the United States should have followed during the past quarter of a century emerge ever more clearly.

First, given that free enterprise must function competitively if it is to make the best use of economic resources, the United States Administration should have resisted and not connived in the prorationing schemes by which the oil producers sought to protect themselves against over-production. Over the years, the result would have been modestly cheaper oil, less wasteful over-drilling of established oil fields and more vigorous exploration of new oil fields, especially those on the continental shelf.

Secondly, the Administration should have developed more rational methods of regulating the natural gas industry. The objective should have been to see that the price of gas from the most expensive gas wells was roughly the same as the price of oil from the most expensive domestic oil wells, calorie for calorie. The simplest way of doing this would have been to abandon overt regulation, but to recoup some of the profits of the no doubt prosperous gas producers by the straightforward operation of the taxation system (with depletion allowances dispensed with).

Thirdly, on imports, the United States should have recognized that, far from being a threat to the domestic industry, imports of cheap oil from the Middle East were a potential strategic benefit, recognizing that its own long-term interests would be best served by using somebody else's petroleum reserves and not its own. But the scale of imports could, to some extent have been controlled by import duties.

Finally, in the several energy-related environmental arguments of the past decade, the United States Administration should have been fierce in defence of the principle that environmental amenities of one kind or another are public purchases which cannot be had for no cost to the economy. The enthusiasm of Congress for cleaner air and water in the years since 1964 might well have been moderated if the administration had done the simple arithmetic to show how much the American population would ultimately have to pay.

Internationally, the consequences of United States policy on energy have been profound. The abrupt re-entry of the United States into the international market for oil in the late 1960s aggravated the difficulties of the other major oil consumers in their relationship with OPEC producers. The fact that the short-term capacity of the United States to increase its domestic supplies had by then vanished meant that the industrialized countries of the world were unable to resist the demands eventually made for increased prices. The limitation on oil imports to the United States, conceived of in 1959 as a measure to preserve the strategic independence of the United States, coupled with entirely inconsistent domestic policies whose effect was to encourage consumption, had become a strategic weakness. The surprise is not that the oil producers were able to exploit the situation to their own advantage, but that they were so slow to do so.

5
From concession to cartel

The Bible has a parable about the wise and the foolish virgins. There were seven of each, all of them due at a wedding and each provided with an oil lamp. The foolish ladies let their lamps burn all night, and then found themselves badly equipped to attend the wedding. The wise put out their lamps at night and lit them again the following day. In a peculiarly uncharitable display of smugness, the wise virgins are reported to have told their sisters that it is foolish to consume today resources you may need tomorrow. The moral is, presumably, 'Waste not, want not.' The question is whether the parable is applicable to the industrialized nations of the West (Japan included), which have allowed themselves to become dependent on petroleum. Are they just another pack of foolish virgins? Or is it the oil-producing countries who, by allowing themselves to be exploited, have become the foolish virgins? The parable, like most parables, is too simple to be taken seriously. Both the oil consumers *and* the oil producers have behaved foolishly. They continue to do so.

The tale of how the oil consumers have come to depend on the oil producers is important in its own right, but is also a necessary means of understanding how the events of the past few years have been set in train. There is, however, a preliminary heresy to be disposed of.

It is quite beside the point to complain that the world has been foolish in allowing its dependence on petroleum to increase so rapidly since the beginning of the century, when petroleum is both the least plentiful of the fossil fuels and a convenient source of petrochemicals. Since there has not been, and is not, any physical shortage of energy, it is entirely sensible for the cheapest fuels to have been exploited first. To have done otherwise would have been literally to waste economic resources. And to complain that petroleum should have been left in the ground during the early decades of this century for the sake of a petrochemicals industry which did not then exist is as unreasonable as it would be to call the British Bronze Age profligate for its use of tin for making alloys with copper without considering the needs of the tinplate industry that would come into being at the end of the nineteenth century.

The question of how best to exploit a natural resource is much discussed. Economists argue that it is prudent to exploit the cheapest resources first. Those who call themselves ecologists ask that thought should be given to future generations. The conflict between these two points of view is not necessarily, to reasonable people, as great as is sometimes made out. First, the economists' argument, when properly qualified, is not a recipe for a fickle and wasteful exploitation of the most easily accessible parts of a natural resource. If, for example, the question arises whether it is wise to replace oil production from an existing well by oil production from one potentially more productive, it is only proper that the capital cost of the new venture should include not merely the cost of drilling holes in the new ground, but also the cost of writing off the capital locked up in the old venture. If the new development will create unemployment, or some other social disbenefit, the corresponding costs should strictly be reckoned in the true (or 'resource') cost of the new enterprise.

With these reservations about the meaning of cheapness, the strategy of 'cheapest first' has the virtue that it helps to create industrial and social institutions which would not otherwise have come into being – in short, it increases the industrial and social capital which future generations have at their disposal. It is certainly mistaken of the conservationists to claim, as they sometimes do, that the interests of future generations will best be served if industrialized societies can somehow restrain themselves from exploiting the cheapest natural resources first, or otherwise fly in the face of the economic arithmetic. In other words, in their wish to use as much cheap petroleum as they could find, the industrialized nations of North America, Europe and Japan have been neither profligate nor indifferent to the needs of future generations.

Where they have been foolish is in their failure properly to appreciate that long-term dependence on imported oil could be made secure only by long-term agreement with the producing countries whose stability would itself depend on the recognition by the producers that they enjoyed a fair share of the benefits of cheap oil. The governments of oil-consuming countries have shamelessly rigged the market, putting short-term and chauvinistic considerations before what should have been their real interests. Thus have the oil producers been given a sense of having been exploited. But now that the balance of economic advantage is reversed, it is the consuming nations that consider themselves affronted. Since the upheaval of the petroleum market in the second half of 1973, the oil-consuming nations have talked endlessly about how the shortage of energy supplies has emphasized the interdependence of the autonomous nations of the world. They are right to do so, and no doubt the OPEC nations will, in due course, discover that they have as much to lose as to gain from oil prices so high that potential customers are driven to spend resources on the development of alternatives. But the oil-consuming states should have recognized much earlier that the rapid growth of the international market in petroleum over the past half century implied a degree of interdependence between them and the oil-producing states that could not safely be ignored.

The essential issue is how should the oil consumers and the oil producers have divided, and even now divide, what is called the economic rent from the exploitation of petroleum? The cost of producing a barrel of oil is a measure of the economic resources expended, and the Middle East stands out as an area where the production cost of oil is low – a mere $0.15 a barrel or thereabouts. The value of a barrel of Middle Eastern oil in Western Europe or North America is numerically the same as the price of the fuel that would have to be used instead if imported oil were not available, and if the market is not artificially constrained, that price is a measure of the cost in economic resources that must be expended to produce the alternative to a barrel of oil. The difference between the economic value and the production cost, allowing for such things as the cost of transport and the interest on the capital employed, is the economic rent. The issue that has arisen between the OPEC nations and their customers is that of how the economic rent should be divided. The roots of this dispute lie in the long history of petroleum exploitation outside the self-sufficient industrialized countries, and the history of oil production in the Middle East in particular.

Although the legendary (and biblical) eternal flames near Kirkuk in Iraq have been known for centuries, the region became a significant oil producer only slowly. Throughout the nineteenth century, the United States was followed principally by Romania and what is now the Soviet Union, and much of the small but quickly growing European demand for petroleum was met from production around the Black Sea and the Caspian as well as from the Far East, chiefly from Sumatra and what is now called Indonesia.

The first exploration licences in the Middle East were let by the government of Persia (now Iran) to the d'Arcy Exploration Company in 1903, and during the First World War became the foundations on which the British Petroleum Company was built. With the collapse of the Turkish Empire during the First World War, the rest of the Middle East became accessible to the exploration companies, at first those of Britain and France and later to American interests.

The pace of exploration, discovery and development was in retrospect slow. American interest was in part diverted by oil discoveries in Venezuela, nearer home, and, in 1930, by the opening up of the East Texas oil wells. In 1944, the proved reserves of the Persian Gulf (Iran, Iraq, Kuwait, Saudi Arabia and Qatar) were estimated at only just over 2000 million tons of petroleum – less than half the petroleum extracted from them to date. Meanwhile, the proved reserves of the Persian Gulf have increased to more than 50 000 million tons.

The reasons for the slow emergence of the Middle East as the dominant force in the international market for petroleum are not far to seek. First, there is the psychological truth that it is easier to appreciate the value of a productive oil well on your doorstep than a much richer deposit of petroleum 3000 or 5000 miles away. Secondly, although the productiveness of individual oil wells around the Persian Gulf was quickly apparent, the geophysical techniques needed to demonstrate the great size of the petroleum reservoirs there had not been developed. Thirdly, there were problems of transport – in the chief market for petroleum, the United States, it would have been unthinkable in the years before the Second World War for any but a small proportion of the demand to be met with oil tankers of the sizes then in service. Fourthly, the market for petroleum in Europe had not then been created – petrol or gasoline consumption was growing quickly, but coal remained the chief source of other forms of energy. Even so, the strategic importance of the Middle East as a source of petroleum was plainly recognized by the Western Allies during the Second World War, both in the military importance they collectively attached to the Middle East and in the care with which British and, less successfully, French interests sought to ensure that future exploitation would rest chiefly with European rather than American companies.

At the beginning of the exploitation of Middle East petroleum, the producing companies were welcomed by their host governments as a potential source of revenue. Even as late as 1938, what became the Aramco concession in Saudi Arabia was let for £50 000 in gold. In the standard agreement between concessionary companies and leasing governments, the companies were given exclusive rights to explore and later develop oil reserves in some designated tract of country, financing the work entirely out of their own resources. The earliest agreements, such as that between British Petroleum (then the Anglo-Persian Oil Company) and the government of Persia, allowed for a fixed payment equivalent to $0.15 for each barrel of oil exported. Later agreements were drawn up on a royalty basis, typically 12½ per cent of the export price, which is historically the origin of the system by which oil companies undertook to 'post' (or publish) prices for crude oil at all times, and in those countries where oil production was determined by royalty agreements. With increasing competition for exploration licences after the Second World War, producer governments were increasingly anxious to include in their agreements with the companies stipulations about the limited tenure of leases, the pace of exploration and development work and the handing back of predetermined proportions of leased territory relatively early in the licence period. In spite of these largely administrative arrangements, however, the standard return to the producer governments until the early 1950s was modest, even meagre, amounting to little more than $0.15 for each barrel of oil brought to the surface. With annual production from the Persian Gulf as a whole running at less than 50 million tons, the total revenues of the governments concerned were approximately $50 million a year.

The upheaval of the past few years is the culmination of a series of changes set in train in 1950. By then, the government of Venezuela had been able to negotiate an agreement with its concessionary oil companies for payments calculated as half the profits attributable to exploitation. By 1950, the American consortium Aramco, with leases in Saudi Arabia, had extended the principle of equal profit-sharing to the Persian Gulf. The government's take could be simply calculated from the posted price – itself, to begin with, the price at which the oil companies would sell to third parties. The Aramco agreement in Saudi Arabia probably helped to bring to a head nationalist sentiments in Iran which were, in the early 1950s, effectively canalized by the Iranian politician Dr Moussadeq.

In retrospect, the dispute between British Petroleum and Iran has a faintly comic air. At a time when oil revenues in the Gulf were increasing, the Iranian Government found that its revenues were decreased by the stipulation in the agreement that its income should be linked with the dividend paid to the company's ordinary shareholders, itself restricted by the then British Government's policy on dividend restraint. A draft agreement to compensate the Iranian Government for this anomaly was never ratified by the Iranian parliament, but countered with a demand for a share of profits. British Petroleum's case for initially declining to extend the principle of profit-sharing rested on the perfectly proper ground that the Iranian concession had been let in 1904 for seventy-five years, and that there was no legal basis for changing these rules. Although eventually the company softened its position, Moussadeq, then prime minister, was able to carry though a decree nationalizing the oil refinery at Abadan, originally built in 1924 and establishing the National Iranian Oil Company to develop and exploit Iranian oil. His success was short-lived, for British Petroleum was quickly able to substitute crude oil from Kuwait for that denied it in Iran while preventing the sale of Iranian oil to third parties by appeal to the international courts. By 1954, however, with the demand for petroleum increased by the Korean War, both parties were able to settle their differences, though British Petroleum would never again be the sole concessionary in Iran, but merely a member of the Iran consortium, with 40 per cent of the original concession.

The events in Iran between 1952 and 1954 were, or should have been, a powerful sign of changes to come. Postwar Iran was understandably turbulent. The country had been occupied rather cursorily by the Western Allies during the Second World War, and the then Shah, pro-German by inclination, had been exiled. Dr Moussadeq's radical nationalism was entirely unfamiliar in the early 1950s, though it was later a common accompaniment to the emergence of nation states in Africa and elsewhere. If Moussadeq had not miscalculated the strength of his hand, and if he had not in particular, underestimated the problems of selling Iranian crude oil on the international markets while crude oil was in surplus, it is possible that the present Shah, exiled for a time during the nationalization crisis, would never have returned.

As events turned out, the Iranian nationalization crisis could have provided both oil companies and oil-consuming governments with a vivid but harmless illustration of how oil can create political problems. For one thing, it showed that mere insistence on the legality of a commercial agreement between an oil company and a government is not a sufficient defence against expropriation or some other unilaterally imposed change of terms. Secondly, it should have demonstrated to all those dependent on Middle East oil that security of supply would depend critically on whether the producing governments could be persuaded that the terms were equitable. Many governments did take fright at what had happened in Iran, and in Britain, fears for the security of oil supplies helped to put wind in the sails of what turned out to be the over-ambitious nuclear power programme of 1955. American companies were not directly affected by what had happened in Iran, but the general sense of insecurity helped to stiffen determination to protect the domestic market. The response of companies and consuming governments alike was, however, entirely negative.

From 1950 to 1974 the posted prices for oil remained the basis for calculating the revenue due to the oil-producing governments. The profit-sharing formula sponsored by Aramco initially required the companies to subtract from the posted price their expenses (production cost plus royalty) and then to divide the difference equally with the government. Even this innovation required that the books of the companies should be open to the inspection of the governments' accountants. The negotiations between OPEC and the companies from the mid-1960s were largely concerned with the amount of the posted price and with variations of the formula for calculating the governments' share: in succession, allowances for the cost of marketing and distribution were abolished; the governments' share was increased, first to 55 per cent, then to over 60 per cent; the admissibility of the royalty payment as an expense was denied; and the posted price was itself indexed against inflation. Until September 1973, these issues were decided by negotiations between OPEC and the companies. Thereafter, they were decided unilaterally by the OPEC members. Given the spread of participation by the governments in oil production, the system became unworkable. During 1974, for example, oil companies were required not only to make payments calculated from the posted price ($7.00 a barrel in Saudi Arabia), but also to buy back from the governments oil not sold elsewhere at a higher price determined only after the event. For perfectly sound reasons, OPEC began, in December 1974, to dismantle the system of posted prices in favour of one that specifies a single price for oil to be sold at variable discounts to the companies with whom the governments have special relationships.

Throughout the 1950s, the posted price of oil was closely related to its market value. It was the price at which oil companies were prepared to sell to purchasers in the Persian Gulf who might carry away tanker-loads to Japan or Europe for use in local refineries. By the mid-1950s, the average payment per barrel of oil in Saudi Arabia was $0.80 per barrel, corresponding to a posted price of $1.72 per barrel. Elsewhere in the Persian Gulf, similar payments became usual. For practical purposes, between 1950 and 1960, the host governments' take multiplied by between three and four while the volume of production often increased still faster.

So why did the producer governments become unhappy with the system? By the late 1950s, it was clear that the supply of oil was likely to grow even more quickly than demand, partly because of the development of the Middle East itself, partly because of new developments in North Africa and Nigeria. But this was the point at which the United States introduced mandatory import regulations. By 1960, it had become plain that Western Europe was destined to become as important a market for petroleum as the United States, though the capacity of the Middle East to supply was growing still faster than the demand in Europe and Japan. In 1960, Aramco in Saudi Arabia was followed by other companies in the Middle East in its unilateral decision to reduce the posted price of oil, thus automatically reducing the revenue per barrel to the producing governments. At the time, the companies argued that it was surely more in the interests of the producing governments for them to retain a share of the international market than for them to maintain their rate of return on each barrel sold at the level to which they had become accustomed. Such arguments cut very little ice with the governments of the Persian Gulf, already alarmed by the effort being spent by the oil companies on exploration in North Africa (principally Libya and Algeria). In any case, the governments had no equity stake in the operations of the oil companies, but were solely dependent for their revenues from oil on a pattern of production essentially determined by the companies and not themselves. OPEC was the child of their anxieties.

From the beginning, OPEC's aims have been ambitious, and today it seems surprising that oil companies and consumer governments were so ready in the 1960s to brand OPEC's initial fumblings as a sign that it could never effectively unite the interests of its disparate members. It is true that OPEC has so far failed to become a fully fledged cartel in the sense of being able to require its members to regulate the volume of their production. But OPEC did win an early success when, in 1962, it secured the consent of the oil companies to a revision of the method of calculating the government's take from the posted price which in effect increased the revenue per barrel above the payments made before the period of unilateral reduction of posted prices in 1960.

It is impossible to guess what would have happened had OPEC never been established. The producing governments had learned from the Iranian crisis of 1952–4 that they could not expect an instinctive and collective reaction among themselves if this implied an overall reduction of output. In 1960, OPEC's immediate objective had been to restore the posted prices for petroleum which the oil companies had reduced. In this task it did not succeed, chiefly because it failed to devise a system or prorationing to which all its members could subscribe. Oil producers such as Saudi Arabia, with the largest reserves, urged that collective action to force up the price by the planned reduction of output should be based on production quotas linked with reserves. Iran, however, asked that a producer's share of the agreed output should be determined by its need of revenue, and thus by its population. The members of OPEC still differ profoundly in what they hope to gain from the future exploitation of their oil reserves. At one end of the spectrum are countries like Algeria, Iran and Venezuela, anxious to use their petroleum reserves as a springboard for industrial and social development. At the other are countries like Saudi Arabia and Kuwait, unable for the present to make full use of their revenues from petroleum in domestic development and fearful that money in the bank, however wisely placed, may eventually be depreciated by economic forces which they cannot influence directly. Given that the members of OPEC also span a wide political spectrum, from the zealous anti-Communism of the Libyan regime to the Baathist government of Iraq, it is no wonder if OPEC has failed to devise a common front on every issue.

Nevertheless, OPEC has much to boast of. Although the organization failed in its declared objective of restoring the posted price of petroleum to the levels set in the mid-1950s, its mere existence prevented the further reductions of posted prices that would have been entirely logical in the early 1960s, when the capacity of the oil companies to produce petroleum exceeded demand. By 1962, OPEC was also able to negotiate with the oil companies a revision of the treatment of royalty payments in the calculation of the payments to producer governments which virtually increased their share of the profits on a barrel from 50 to almost 60 per cent.

Throughout the 1960s, OPEC was, as an organization, chiefly occupied in the closer definition of its common objectives, while individual member governments were separately occupied in attempts to persuade the oil companies to increase the scale of their operations, and thus of the oil revenues payable. The clearest statement of OPEC's common interest was published as a communiqué in 1968. On 25 June, OPEC laid down the principles that producing governments should have the right to explore and develop their own oil fields to the 'maximum extent possible', and that the participation of outside interests, principally the oil companies, was to be undertaken on terms which could be revised periodically. Host governments were to be free to acquire a share in the ownership of companies operating concessions 'on the grounds of the principle of changing circumstances'. Further, they would be free to ensure that concessionary companies could not obtain 'excessively high net earnings after taxes' and would be entitled to recover the excess above 'the reasonable expectation that would have been sufficient to induce the operator to take the entrepreneurial risks necessary', if necessary by retrospective imposts.

In short, OPEC has, for the past decade, made no secret of its belief that its members' best interests demand a total control over the scale of the companies' operations, a high and increasing degree of participation in those operations and the right to ensure that such benefits as accrue should find their way back to the producing governments. The understanding that revenues should be calculated from published posted prices has, in the past, given each OPEC member an assurance that no other will seek a larger share of the international market in petroleum by unilateral price reductions. Within this framework, OPEC members have remained free to pursue whatever policies they think necessary in their own interests.

Now it is proposed that the cumbersome but explicit formula based on the posted price should be abandoned. If OPEC governments become able to make their own commercial deals with those who buy their oil, and, in particular, are not required to make public the discounts they offer to individual oil companies, individual oil producers needing increased revenues will be more able than in the past to undercut their fellows. In short, the abandonment of the system of posted prices will be a welcome step towards the creation of a free international market in crude petroleum, though it will provide a potential source of conflict within OPEC.

During the 1960s, the market for petroleum products outside the United States was weak: the price of crude oil and petroleum products of all kinds declined in Europe by roughly 15 per cent, chiefly because of the rapid increase of production capacity in Africa and the Middle East, the introduction of 200,000-ton tankers and the steady improvement of petroleum technology in refineries. In the United States, virtually isolated from the rest of the world's petroleum market, prices moved differently – the small declines in the first half of the decade were more than made good by the increases which set in after 1965 and which were, in part at least, a consequence of the higher tanker charter rates occasioned by the rapidly growing demand for heavy fuel oil on the east coast of the United States.

The meeting of OPEC in June 1968 is, in retrospect, the turning point in the relationship between oil producers and their customers. Although the meeting failed once more to yield an agreed formula for regulating production from OPEC oil fields, individual countries had accumulated a great deal of evidence that the oil companies operating OPEC concessions were in no position to resist quite substantial demands. In the mid-1960s, the Algerian government had been able to increase its revenue from Algerian oil sold directly to the French state-owned oil companies above the levels originally agreed as part of the treaty signed in 1963 to provide for Algerian independence. The French concessions in Saudi Arabia had also been let on terms which gave the Saudi Arabian government's own national oil company a recognized share of the output to sell independently. Elsewhere in the

Middle East, national oil companies in Iran and Kuwait had asked for and obtained supplies of crude oil from their concessionaires at favourable prices. The government of Iraq, which in 1963 had unilaterally cancelled exploration rights in parts of the country not already producing oil, had found the oil companies only too eager to renegotiate their leases.

At the same time, it had become clear that the established multinational oil companies, and in particular the eight 'majors', were no longer the only companies willing to enter into lasting agreements. Apart from companies like American Occidental Petroleum, which had sprung to prominence in the early 1960s by its successful exploration in Libya, there was a steady growth in the numbers of 'independents' from North America and Japan willing either to lease concessions on the old-fashioned basis or to buy crude oil from the publicly owned oil companies of the producing states. Although prices were still falling, the scope for selling crude oil seemed greater than ever.

It was thus inevitable that the participation of the producing governments in the commercial equity of the oil companies should have become an open issue. Some governments, like that of Iran, had made no secret of their wish to be more intimately concerned in the processing of crude oil, partly for the industrial benefits and partly because they had good reason to think they they could capture a still greater share of the economic rent of crude oil if they were engaged in the 'downstream' activities of the oil industry – refining, distribution and even marketing. Other OPEC members were moved by less tangible considerations: their resentment of the high profits of the oil companies and at the size of the excise taxes levied in European countries, and their fear that inflation would undermine the value of their revenues.

Soon after the OPEC meeting of 1968, Sheikh Ahmad Yamani, oil minister of Saudi Arabia, said it was his government's intention to 'establish control over all oil operations . . . The major part of our long-term plan will be attained in ten years . . . it may take up to twenty-five years to attain full control.' This declaration was echoed by Sheikh Yamani's opposite number in Kuwait, who added the ominous warning that it would always be possible for a producing government to increase the taxes on crude oil to the level at which an oil company's concession would yield only a small profit, whereupon the company could be nationalized at comparatively little cost. In the event, the long-established pattern of the relationship between producing governments and oil companies was broken not in the Middle East but in Libya.

In September 1969, King Idris of Libya was overthrown and replaced by a government headed by Lieutenant (later Colonel) Ghadafi. Six months later, in May 1970, the Libyan government ordered a reduction in the output of its oil wells when the market for crude oil from the Mediterranean was already straitened by the cutting of the trans-Arabian pipeline in Syria, an event reported as an accident but which was indisputably a collusive act. One immediate consequence was a dramatic increase of the charter rates for tankers, which served quickly to strengthen the Libyan demand for higher taxes on crude oil. (In the complicated partition of oil revenues between governments and the concessionary companies, the Libyan posted price is traditionally higher than that in the Persian Gulf, not only because Libyan oil is low in sulphur, but also because transport costs to markets in Europe and America are less.)

Much of the Libyan case for higher taxes was based on the artificially inflated tanker rates. The resistance of the oil companies was partly based on the accurate calculations that a concession to Libya at that stage would be followed by leapfrogging demands from the states around the Persian Gulf once the tanker market returned to normal. In the end, the negotiations between Libya and Exxon, the concessionary company singled out as the one first to conclude an agreement, were circumvented by an agreement between the government and Occidental that the posted price of Libyan crude should be immediately increased by $0.30 a barrel, with further increases at annual intervals.

This was the first time that a producing government had been able to play off oil companies against each other after the letting of exploitation leases, and by October 1970 the other main producers in Libya (British Petroleum and Shell as well as Exxon) had been forced to follow suit. Fears that increased prices for Libyan crude would be followed by demands for higher prices in the Persian Gulf were quickly justified, and the posted price of Iranian crude was increased by $0.16 a barrel before the year was out. The oil companies recognized that the old order, in which for practical purposes it was they who fixed the posted prices for oil, was gone for good.

Anxiety about the effect of the Libyan agreements on the posted price of oil in the Persian Gulf explains why the oil companies were almost eager to take part in full-scale negotiations with the six Middle East producing governments (Abu Dhabi, Iran, Iraq, Kuwait, Qatar and Saudi Arabia) in the early weeks of 1971. The Tehran Agreement, signed in February 1971, was regarded by the companies as a settlement that would provide them with stability for at least five years. The essence of the agreement was that the posted price of oil should be increased immediately by $0.35 a barrel, and that there should be annual increases of 2.5 per cent to take account of inflation in the industrialized West. The Gulf States were to be compensated for any failure by the oil companies to match the posted price of Libyan crude oil with the current tanker rates, but otherwise undertook not to base arguments for increases of their own posted prices on price anomalies elsewhere.

Although the Tehran Agreement finally abolished the minor allowances the oil companies had enjoyed to cover the cost of freight and marketing and also placed the calculation of the government's revenue on a uniform basis of 55 per cent of the difference between posted price and the sum of production cost and royalty, the companies (like the United States government) were grateful for what appeared to be the promise of stability. Within weeks, a similar agreement had been signed with the remaining OPEC members at Tripoli, and although this was modified in 1972 to take account of the devaluation of the dollar, there was a widespread feeling in the oil industry and among its customers that at least they knew the worst that the succeeding five years could bring. Events, of course, proved them wrong, and this much should have been apparent.

The spectacular weakness of the Tehran and Tripoli Agreements was that they said nothing about what had by then become an increasingly insistent theme: the demand for equity participation by producing governments in their own oil production. Strictly speaking, the question of the price of crude oil with which the agreements dealt is logically distinct from the question of how great a share the oil-producing governments should have in the exploitation of their oil resources. In fact, it should have been plain that the producing governments would not consider themselves bound by the Tehran and Tripoli Agreements without intensifying their interest in participation. The governments had, after all, made clear their intention of winning for themselves as great a share as possible of the economic rent of their crude oil, and even after the Tehran Agreement, there was still a long way to go. If the government take was to be stabilized for five years, the next most obvious strategy would be to acquire larger amounts of crude oil for sale direct on the international market.

So much is clear from the movement of the posted prices and the taxation rules in force in 1970. In the early months of that year, for example, the posted price of the standard Saudi Arabian crude oil was $1.80 a barrel. With a 12.5 per cent royalty (the equivalent of $0.255 a barrel) and allowed production costs of $0.13 a barrel, the government's share of the national profit on a barrel of oil was $0.663 (half the difference between the posted price and the sum of royalty and production cost). The result was that the cost of the oil to the companies was $1.038 a barrel (the sum of production cost, royalty and tax) and the government's revenue was $0.918 a barrel. In the wake of the Libyan settlement, the tax rate was increased to 55 per cent, while the posted price was increased to $2.18 a barrel under the terms of the Tehran Agreement. What this implies was that the cost of oil to the companies was increased from $1.038 in the early months of 1970 to $1.376 in February 1974, with the prospect that the escalation clauses of the Tehran Agreement would carry the cost to $1.639 in 1975. During the same period, the Saudi Arabian government's revenue on a barrel of oil was increased from $0.92 in early 1970 to $1.256 in February 1971, and was destined to increase year by year to $1.519 in 1975.

Substantial though these increases, real and prospective, were, the producing governments were not unaware that their own revenues were less than those collected by consumer governments from taxes on petroleum products – reckoned by the OECD to average more than $5.00 an imported barrel of crude oil. Although the bulk of this taxation was intended by the consumer governments as a device for recovering part of the cost of building roads and backing the development of road transport, it was inevitable for the producing governments to take the disparity between their own revenues and those of the consuming governments as a sign that the market could shoulder a bigger burden. And the best way of doing this within the framework of the Tehran Agreement would be to acquire a share of the oil produced and to sell it for as much as it would fetch.

The Tehran and Tripoli Agreements also demonstrated to the OPEC governments that the oil companies were no longer able to present a united front. The reasons were plain. In spite of the rate at which money was being spent on exploration outside the OPEC countries, there was no prospect of the oil companies being able to supply the demand for petroleum in Europe, Japan and the United States without supplies from OPEC producers. Worse still, for the companies, the cost of crude oil from the Middle East was still, even in 1971, less than the likely cost of crude oil from other potential reservoirs, like the North Sea and Alaska. So the chances were high that the companies would grin and bear whatever further demands the concessionary governments made.

The predictable collapse of the Tehran and Tripoli Agreements came quickly. From the beginning of 1972, participation was in the air, and Aramco in Saudi Arabia was the first of the oil companies to agree that there should be a ten-year transition period in which 51 per cent of its equity would be transferred to the national petroleum company. Similar agreements were reached elsewhere in the Gulf (with the notable exception of Iran). In 1972, the Libyan government expropriated British Petroleum's share of a concession, on political grounds, and also asked for an immediate 51 per cent participation in the affairs of its concessionary companies. Occidental agreed, and the reluctance of the major oil companies in Libya was overridden by decree early in 1973. Early in 1973 at Riyadh, the capital of Saudi Arabia, OPEC accepted a general formula for the transfer of ownership of its oil concessions so that all producing governments would own 51 per cent of their oil business by 1985 – a formula since overtaken by events.

The workings of participation are obscure, and will no doubt remain so. Theory is only a poor guide to practice, for the relationship between a participating government and an oil company is not that between a shareholder and a company in which he has acquired a financial stake. First, the financial basis on which OPEC governments have acquired stakes in their oil companies is open to endless argument. The pattern set in Libya is that the value of a local oil company should be set at the depreciated value of its past capital expenditure, or what the accountants call 'net book value', though this makes no allowance either for the extent to which a company's expertise has made possible its success nor for the potential profitability of its concession (and which the multinational company behind the local operation may have used as an argument for raising capital investment funds). Secondly, the producing government usually pays for its stake with oil, not cash. Thirdly, the producing governments have done their best to ensure that agreements on participation require the oil companies to buy back whatever oil the government is unable to sell at prices which are a substantial fraction (roughly 95 per cent) of the posted price. Not merely are the oil companies tax-gatherers for the producing governments, but they are obligatory commission agents as well.

Once the principle of participation had been conceded by the end of 1972, the downfall of the Tehran and Tripoli Agreements was assured. The producing governments soon discovered that they were able to see substantial amounts of their share of the crude oil from participation agreements at prices much greater than their revenues from royalties and taxes, a discovery made easier in the tight winter market of early 1973, when output from Libya and Kuwait was reduced by government decree. By the spring of 1973, the Shah of Iran had startled the members of the Iran Consortium with a declaration that his government would immediately take full control of its oil production. The arrangement, ratified by the Iranian parliament in July 1973, reduced the technical part of the consortium to a contractor for the Iranian National Oil

Company, but the agreement on the disposal of oil allows the companies belonging to the consortium supplies of oil at prices roughly the same as those which other partners in participation agreements enjoy. In many ways, as will be seen, this apparently unfavourable arrangement, in which the Iran Consortium was stripped of its right to decide how the concession should be managed, turns out to have some unexpected benefits.

By the summer of 1973, just over two years after the Tehran Agreement had been signed, OPEC asked for renegotiation on the grounds that its members' revenues were inequitably low. Negotiations began, and were promptly adjourned to allow the oil companies a chance to decide how much extra they could afford to pay. The meeting never reassembled, for on 16 October 1973 OPEC announced new posted prices for crude oil which had not been negotiated with the companies but fixed unilaterally. The Saudi Arabian posted price, already increased, as foreshadowed by the Tehran Agreement (to allow for the devaluation of the dollar), from $2.39 to $3.01 a barrel, was increased by 70 per cent to $5.12 a barrel. In the closing days of December 1973, the posted price of oil was, for practical purposes, doubled, with the intention of providing Saudia Arabia with an income of $7.00 on each barrel of oil exported by the companies and with correspondingly more from its own share of the output of its oil fields. Within five years, the income of the oil consuming states on each barrel of oil sold had multiplied eight times.

The culmination of these changes on price and participation coincided with the Arab–Israeli war of 1974, and with the oil embargo of most (but not all) of the Arab members of OPEC against some oil consumers, the United States and the Netherlands especially. The war had already begun when OPEC and the oil companies met for their last and abortive round of joint negotiations, but the meeting was adjourned for two weeks. Later in October, OPEC announced unilaterally from Kuwait that prices would be virtually doubled, and the following day the Arab members of OPEC announced that October's production would be restricted to 95 per cent of that of September. The following month, the output restriction was increased to 25 per cent, with the promise of a further monthly 5 per cent reduction until the Israeli question had been settled. In practice, Iraq declined to endorse this proposal on the grounds that it was not possible to discriminate between friendly and unfriendly customers, a view borne out by the way in which the international oil companies were able to share such supplies as were available more equitably among their customers. Libya, though formally one of the Arab states committed to the notion of an embargo, appears to have been less than meticulous in its enforcement.

The effectiveness of the 'oil weapon' as an instrument of political as distinct from commercial policy is unclear. By the end of 1973, the Arab states had wrung from Japan and the members of the European Community ambiguous declarations of sympathy with the Arab view that a settlement of the Israeli question, and in particular a revision of the boundaries of Israel established after the 1967 war, was necessary. At the same time, the effect of the restrictions on output was diminished by the willingness of non-Arab OPEC members, Iran in particular, to increase production, and by the recognition in the closing weeks of 1973 that the output restrictions could cause economic difficulties in Europe and Japan without substantially influencing United States support for Israel. The oil weapon may have helped to colour the eventual truce, but there is nothing to suggest that the United States did more than accept that a permanent settlement in the Middle East is necessary.

Throughout this period, the position of Saudi Arabia was all-important. The first suggestions that Saudi Arabia might abandon its traditional view that oil supplies should not be used for political purposes came in August 1973, but in less than a year Saudi Arabia had abandoned this policy in three important respects: (a) by agreeing that the embargo should be lifted even before the United States had taken the initiative in bringing about negotiations between Egypt and Israel; (b) by making plain its view that the well-being of the oil producers was linked with that of the oil consumers and urging that the price of oil had been pitched too high at the end of 1973) and (c) by making arrangements that production capacity as distinct from production in Saudi Arabia should be rapidly increased. Because Saudi Arabia has larger reserves of oil than any other OPEC member, and can now if it chooses increase the scale of its production quickly, its influence on petroleum prices and supply policies in the years ahead is bound to be decisive.

If the political consequences of the first use of the 'oil weapon' were inconclusive, there is no doubt that the reduction of output towards the end of 1973 helped to establish the prices decreed effective from the beginning of 1974. With Japanese, European and American independent oil companies bidding for OPEC oil at close to the posted price, it would obviously have been impossible for the oil-consuming states effectively to establish the view that prices had become unreasonably high. And since the first quarter of 1974, there is no evidence that oil in the Persian Gulf has changed hands at prices in excess of $12.00 a barrel.

The consequences of the transformation of the petroleum market are far-reaching. They are the subject of much that follows. But several more immediate and recriminatory questions arise. Has OPEC gone too far? What, in any case, is the future for OPEC? And has the role of the great multinational companies as intermediaries between OPEC and the ultimate consumers of petroleum been a blessing or a curse?

The first thing to acknowledge is that OPEC has indeed succeeded in its declared intention of 1968 in capturing the lion's share of the economic rent in crude petroleum. Whether they have captured all of it, or have even overshot the mark, only time will tell. It is easy enough for an OPEC government to calculate the cost of producing a single barrel of oil. The uncertainty is in knowing just what source of energy that barrel will displace when it eventually finds its way to its consumers. During 1974, the market price of fuel oil in North America and Europe fluctuated narrowly between $10 and $11 a barrel, corresponding in terms of energy content to rather more than $50 (or £20) for a ton of coal. If the alternative to importing petroleum from OPEC countries were to obtain electricity from nuclear power stations, it would be economically worthwhile to have paid anything up to $700 for a kilowatt of generating capacity, substantially more than the cost of actually building a nuclear power station. If the circumstances which now obtain persist long enough to allow new coal mines to be sunk or nuclear power stations to be built, then OPEC members will have overstated their case for a fair share of the economic rent. Two immediate conclusions follow. First, the prices now being charged for OPEC oil cannot indefinitely be sustained unless the producing governments are willing over time to reduce output substantially. Secondly, the speed with which the facts of life will be apparent to all concerned, but especially to OPEC, will depend critically on the speed with which the principal consumers of OPEC oil can bring into service alternative sources of energy or reduce their imports of oil. Speed is what matters, though it would be hard to infer this from the responses of the industrialized nations to the events of the second half of 1973 (see below).

OPEC has been much despised, but wrongly. Most probably, its members are as vividly aware that it cannot last as its critics. Sooner or later, oil-rich Saudi Arabia and oil-poor Iran will go their separate ways. But that will be at the end of the century. For the time being, they have more in common than the quarrels that divide them. So much is clear from OPEC's comportment. It has worked pragmatically if deliberately. The member nations have tried to organize a system of prorationing and have failed, but none of them despairs, even if it follows that OPEC will never be the first fully fledged international governmental cartel. OPEC has advertised its objectives well in advance, and has shown an admirable, or at least enviable, capacity to wait until the desirable became practicable. In exactly the same way, nobody should be surprised if the time should come when OPEC is no more than a relic of its present self, an international club that gathers occasionally to wonder at the days when a government could reckon to collect $10.00 net (in 1974 prices) for each barrel of oil exported at the expense of the oil companies.

Among the several reasons why OPEC cannot last, a few stand out. The racial and political diversity of its membership is, in the long run, less important than its members' economic and geological diversity. That Iran is not a strict Moslem state and that Iran and Iraq are at daggers drawn about Kurdistan at the north of their common border and about the neutral zone of Kuwait in the south has not prevented them from agreeing that $7.00 a barrel or thereabouts is a sensible tax to put on oil in the circumstances of 1974. But, as the years go by, Iran and Iraq, not to mention Saudi Arabia, will become very different countries. Even if Iran is able in the decade ahead to extend its reserves (at present reckoned to be roughly thirty years of supply at the present rate of output of roughly 300 million tons a year) by an amount equal to the oil consumed in the interval, it will be differently placed from the oil-producing states further south in the Persian Gulf. Iran will be less well blessed with its own oil, and a larger consumer of it, especially of the industrial development programme on which it is now embarked succeeds.

Different OPEC members react differently to the prospect of prices being kept at the new levels for too long. Iran, anxious to gather income in the short run, is understandably concerned that prices should be high in the coming decade. Saudi Arabia, with potentially much larger reserves of oil is more anxious to have a long-term outlet for crude oil in large quantities. At the OPEC meeting in Vienna in September 1974, there were tangible signs of this conflict at work, for Saudi Arabia declined to follow an OPEC decision that the royalty rate on oil companies' own oil should be increased from 14.5 per cent (fixed in July 1974) to 16.66 per cent and that the tax rate should be increased from 55 per cent to 65 per cent. To forgo in this way an increase of taxes likely to add between 3 and 5 per cent to government revenues does not, of course, amount to a serious break in the façade of OPEC, but is a sign of things to come.

The role of the oil companies, and especially of the so-called 'majors', is also likely to be substantially changed. The terms on which oil concessions have traditionally been let have allowed the companies full control of the production and marketing of oil, but not since 1960 have the companies been free agents. From then on they were no longer fully in charge of the scale of production, while the Tehran and Tripoli Agreements in 1971 turned them into tax gatherers for the OPEC governments, and the events of 1973 robbed them of an effective voice in deciding the level of OPEC taxes. Participation has still further eroded their independence, especially because of those provisions of most participation agreements which require the companies to buy back from the governments oil that would otherwise go unsold.

The absorbing question is how far the castration of the oil companies will go. Although it is customary for the oil producers as well as the public in oil-consuming states to regard multinational oil companies as nefarious middlemen making unreasonable profits, the chances are that their existence as intermediaries is necessary for both sides. They were able to soften the effects of the Arab producers' boycott on supplies of oil to the Netherlands during November 1973 and succeeding months; and hence the flexibility of their distribution system may provide is benefit without which consumers would be less comfortably placed. But the oil companies also serve an indispensable function for the oil-producing states, both as an obvious if over-neglected source of technical expertise and as a means of ensuring that the oil-producing governments do not have to stand the whole risk of selling oil in what are almost certain to be, in the next few years, commercially hazardous conditions.

This is probably another sense in which OPEC has probably gone too far for its own good. To the extent that the oil companies in some OPEC countries – Kuwait and Saudi Arabia, for example – are being converted into sales agents selling crude oil on commission, they will find it prudent to direct their attention elsewhere. In many of the oil leases in the Middle East let during 1974, oil companies have been recompensed for their expenditure on exploitation and development by the promise of as little as 15 per cent of the oil eventually produced (to be taxed by way of royalty and a proportion of the notional profit on sales which cannot at this stage be determined). In Libya, new concessions put up during 1974 often required that companies should spend money nor merely on the development of potentially promising oil reservoirs, but also on exploration of less promising parts of the desert. Not surprisingly, the major oil companies have hardly been falling over themselves to make bids.

To suggest that OPEC has in several ways overplayed its hand, and that it is likely to be less powerful in future, is not to scorn what it has accomplished. Its objective, to capture the lion's share of the economic rent, has been achieved, perhaps too much so. Now, like any alliance formed to fight a war, even, as in this case, a war with logic and some justice to commend it, success is the signal for a loosening of the alliance. We may expect to see the initiative passing from the oil producers to the oil consumers.

6
Customers in trouble

If OPEC's successful campaign for a fair share of the economic rent from crude oil has been consistent, logical and deliberate, the response of the oil-consuming states has been uncertain, confused and ambivalent. Although the issues that must in due course be resolved between the oil consumers and the oil producers have been confused by other economic and political problems, it is hard, even in retrospect, to appreciate why the consumers have apparently been so slow to appreciate the change in their relations with the oil producers. That many of them greeted the prospect of a sharp increase of oil prices with the surprised indignation typified by the reference, in September 1973, of Mr George Schultz, then Secretary of the United States Treasury, to the 'swagger of the Arabs', is a measure of their failure properly to digest explicit warnings, often from their own officials.

OPEC's intentions had been explicit at least since 1968. There had been a wealth of objective evidence that the time was fast approaching when relations between the oil exporters and the oil importers would be transformed. By 1968, it was already clear in the United States that it would be impracticable to persist with an energy policy in which imports of crude oil from outside the Western Hemisphere played no part. Indeed, the United States had already become dependent on imports of residual fuel oil (the heaviest fraction of the output of oil refineries), much of it shipped from refineries in Europe or the Caribbean. The reasons were simple, and chiefly economic. Proved reserves of petroleum in the United States were already declining in terms of the steadily growing annual consumption of oil, while the price of fuel oil on the international markets was so low that the electricity utilities in the United States could not lightly ignore the economic penalties they would impose on themselves by counting instead on an increase of domestic fuel oil production. In other words, what should always have been clear had become apparent: there are natural limits to the success of any economic policy founded on the assumption that it is possible not only to conceal but to abolish competitive forces.

If the United States had been consistent in its belief that dependence on imported crude oil from OPEC countries (Venezuela excepted) would be strategically dangerous, it would have set in train policies intended to encourage the development of alternative domestic sources of energy nearly a decade earlier. The result, no doubt, would have been still further increases of the domestic price of petroleum, but probably also a more economical use of energy in all forms. Instead, the administration appointed a Presidential Task Force to examine the consequences of a relaxation of the policy that crude oil should not be imported from outside the Western Hemisphere. This reported, in 1970, that the scale of imports was likely to be only modest and that the strategic risks were small. What the task force did not consider was that the entry of the United States into the international petroleum market would certainly have a profound influence on the price of petroleum and on the terms on which it was obtainable.

The oil companies were more perceptive. By the early 1970s, OPEC's message of 1968 had begun to sink home. In the United States as in Europe it became habitual for oil company chairmen to tell their shareholders about their diminished status in OPEC countries and their anxiety about the continuity of supply. Many government officials were convinced. In the United States, Mr James Akins, the State Department's expert on petroleum policy (appointed American ambassador to Saudi Arabia in 1974) was one of the chief prophets of impending trouble. His main concern was the prospect that OPEC countries would soon be earning financial surpluses large enough to increase their international political standing; his chief mistake was that he underestimated the size of the surpluses, now accumulating by a factor of five. In

Europe, both the European Commission and the OECD (representing Japan as well as the United States and Canada) were increasingly anxious about the continuity of supply, and predicted that the 1970s would be a decade of rising prices. Nobody, however, appears to have appreciated the bargaining strength that OPEC had built up, or the scale and speed with which price increases were impending.

Tangible evidence of the vulnerability of the energy economy of the oil consumers had, however, already accumulated. In the first of several 'energy messages' to Congress in April 1971, President Nixon laid down the long-term policy objective of a return to self-sufficiency, but for two years took no effective steps to attain this impracticable ideal. In the winter of 1972–3, the decision of Kuwait to reduce its output of crude oil, ostensibly on the grounds of the conservation of supplies, but no doubt with a wish to test the sensitivity of the market, was largely responsible for a shortage of heating oil in the United States.

In retrospect, the most charitable explanation of the apparent indifference of governments in the industrialized West to the process which culminated at the end of 1973 is that they had privately acknowledged that the 1970s would be a decade of rising prices and growing participation, but were afraid to admit, even to themselves, and certainly not to their electors, how serious could be the consequences. Only this can account for the way in which the Tehran and Tripoli Agreements of 1972 were welcomed by both OECD and the United States State Department as a framework of stability in the 1970s.

The most immediate consequences of the upheaval in the petroleum market have been financial. The facts are clear, even if neither oil producers nor oil consumers know what should be done about them. In the far-off days of 1965, the total revenues from oil of the OPEC Members amounted to $3847 million. Only Venezuela earned more than $1000 million in that year. Kuwait earned $671 million, Saudi Arabia and Iran not quite as much. In 1972, the first full year after the Tehran and Tripoli Agreements, oil revenues had increased fourfold (partly because of higher production) to $14 700 million, with Saudi Arabia and Iran at the top of the league with $3100 and $2400 million respectively. In 1974, the revenue of the OPEC countries is estimated to be $86 000 million, with Venezuela earning $10 000 million, Saudi Arabia nearly twice as much and Iran nearly as much. By any standards, these are very large sums, and they are likely to increase, whatever happens to the posted price of oil, as participation agreements exert their full effects. Even if the price of oil in the Persian Gulf were reduced to a point where the average revenue of the oil producers was something like $7.00 a barrel, making OPEC oil marginally competitive with other sources of energy, their collective revenue at present production rates would exceed $75 000 million a year.

The characteristic and perplexing feature of these huge revenues is that most of the countries to which they flow are unable collectively to spend all they earn on imports from the oil consumers, industrialized and otherwise. Inevitably, the OPEC countries differ among themselves in the extent to which they can spend what they earn on current consumption of goods and services. Thus while Iran was thought to have earned nearly $4000 million in 1973 from royalties and taxes on oil alone, it was still unable to finance ambitious plans for economic and social development and found it necessary (and easy) to borrow funds from abroad, especially from commercial banks in Europe. When Iranian receipts from oil rose to nearly $20 000 million in 1974, the deficit became a surplus.

Towards the end of 1974, Iran modified its fifth Five-Year Development Plan so as to increase publicly financed social and economic expenditure to $45 000 million in the period 1973–8. The objective is to increase the country's GNP at the rate of 26 per cent a year, which is an unprecedently rapid rate of growth unmatched even by Japan in the 1960s. Given the shown competence of the Economic Planning Agency in Tehran, this target may not be entirely beyond reach, though the most serious bottleneck is the educational system. The result of such growth may, in 1978, be a GNP of $2500 a head of population then exceeding 40 million, essentially matching the GNP of Western Europe. But by then Iran could easily have accumulated foreign reserves and overseas investments of $30 000 million, allowing for probable extra military purchases from abroad. In short, the chances are that oil revenues will enable Iran to join the ranks of the developed countries in no less than half a decade. Sociologists elsewhere, and politicians in Iran, understandably ask whether so much can be accomplished without upsetting the present social order.

The most serious financial problems are occasioned by the revenues of those oil-producing states which are, for the time being at least, unable to make full use of their revenues. Saudi Arabia, as the richest, was budgeting for $27 000 million in the year from 20 July 1974. Of this, the government plans to spend $13 000 million (including $1300 million as foreign aid for friendly states), but whether it can do so is an open question. Even so, it is certain to have $14 000 million to be carried over to future years. This is, in itself, a substantial amount of cash, comparable in size with the annual increase of world liquidity (the scale of the Eurodollar market, for example) in the early 1970s. It is true that, in the long run, the Saudi Arabian surplus may not be so gigantic. Sharing the present oil revenues among the population of eight and a half million people would yield a GNP rather less than the GNP per head of the United States and many European countries (Sweden, for example).

There is, unfortunately, no prospect that industrial developments like those of Iran, Algeria and Venezuela will, in the foreseeable future, consume the present oil surpluses of oil producers such as Saudi Arabia, Kuwait, Abu Dhabi and Qatar. Kuwait, with a population of three quarters of a million, has oil revenues in excess of $10 000 million a year. In the year from 1 April 1974, ordinary public expenditure was reckoned at $1300 million, a little more being set aside for foreign aid to other Arab states (through the Kuwait Fund for Arab Economic Development) and $600 million for long-term investment in local industry and abroad. In other words, between $7000 million and $8000 million will have to find at least a temporary home with international banking institutions. In aggregate, the unspent surpluses of OPEC countries are likely in the years ahead to be between five and eight times the combined surplus revenues of the industrialized nations which belonged to OECD in 1972 before the Tehran Agreements were signed.

The immediate consequences of this transformation of the

international monetary scene have been disconcerting, both to industrialized countries and developing countries like India. There are, in fact, several strategies that the oil-surplus states might follow. To start with, they could spend their surpluses on the import of goods and services. Though this would entail a transfer of actual resources from the oil consumers to the oil producers, it would account for rather less than a year's normal economic growth and would, in the long run, be good for international trade. Unfortunately it will be many years before OPEC countries can follow such a policy. An almost equally desirable course of action – from the point of view of the oil consumers – would be for the surplus revenues to be used for aid to the developing countries without oil of their own. That would have the advantage of allowing the oil consumers to pay for their oil with goods and services (this time, supplied to the developing countries) while relieving the industrialized world of its responsibility to increase the scale of foreign aid and transferring what is essentially a debtor–creditor relationship to the relationship between the oil producers and the developing world. In reality, the oil-surplus countries (especially Iran, Kuwait and Saudi Arabia) committed close on $5000 million to foreign aid during 1974, but, unfortunately for the oil consumers, also insisted that so long as the GNP per head in the industrialized world was greater than that in oil-producing countries, they saw no reason why responsibility for foreign aid should be transferred to them.

Another course that the oil producers might have followed – the investment of their surplus revenues in industrial and commercial enterprises overseas – is also, on the present showing, unlikely to absorb a large part of their surplus cash. On the face of things, the most obvious outlet for such investment funds would be in so-called 'downstream' investment in oil production and distribution. The experience of 1974, as of earlier years, has shown clearly enough that the oil-surplus states are entirely prepared to invest in oil refineries and in petrochemicals plants in their own territories. Indeed, if all the plans for developments like these which have been announced in recent years are ultimately fulfilled, there will be a vast surplus of many petrochemical products in the 1980s. States such as Kuwait are more concerned with tanker construction (on the face of things, entirely logical for a state with very large surpluses lacking the population that would be needed to man land-based industries), but this is a more hazardous enterprise when there is a prospect that the years ahead will see too many tankers chasing too few cargoes. A more telling objection is that investment in tankers on a large scale would inhibit oil producers from limiting the production of oil. But the oil-consuming states, like the international oil companies, have traditionally been anxious to avoid a situation in which a substantial part of the world's tanker fleet was controlled by the countries that produced the oil.

In many ways, the neatest of all the ways in which the surplus funds might be invested would be in the exploitation of alternative sources of energy: oil from elsewhere in the world or even North America, oil shale and tar sands and even the building of nuclear power stations. In the early 1970s, the oil companies alone were expecting to invest something like $400 000 million in the remainder of the decade, most of it outside the Middle East. But the oil producers have no commercial interest in fostering developments that would expose their own products to more intensive competition. Their present oil surpluses exist only because there are at present no easy substitutes for OPEC oil, and it would require an act of unprecedented altruism for them to conspire in the liquidation of their chief commercial asset. In any case, they would be unlikely to give hostages to fortune by providing non-OPEC members with assets that could be sequestered at net book value.

In all the circumstances, it is surprising but significant that, in the past few years, some oil-producing states have been willing to make substantial downstream investments. Iran has been the chief adventurer, but its attempt to acquire a retail distribution for petroleum products in the United States was frustrated in August 1974, allegedly on anti-trust grounds. An earlier scheme for building a refinery in Belgium led, early in 1974, to the downfall of the Belgian government, which had moved with unreasonable and even unpardonable caution in its assessment of the political implications of the project.

Investment in other kinds of industrial enterprises, overseas or domestic, should suit both partners better. The advantages have been apparent for several years. In 1973, Mr James Akins of the United States State Department was arguing the need to encourage 'investments in Europe (and possibly in Japan) as well as in the United States'. Governor John Love, for a brief spell President Nixon's Commissioner for Energy, argued at the same time that it was for the oil consumers to devise ways of making overseas investment attractive to the oil producers. OPEC spokesmen have been divided on the subject. On such issues, OPEC members are certain to differ among themselves, which is no bad thing for the oil consumers. It is therefore more than merely a disappointment that so little has been done, in the oil-consuming states to provide the surplus countries with opportunities for productive investment elsewhere.

Thus it was inevitable that the surplus oil revenues of OPEC states should, after early 1974, accumulate as short-term deposits with the commercial banks in Europe and North America. To begin with, the received wisdom of the banking community in the West was that the growth of these funds need not be disastrous and that the international banking mechanism would be able to 'recycle' them. If, for example, the oil producers chose to place their surplus revenues with banks in Germany or the United States, it would then be perfectly straightforward for those lucky recipients to relend the money to banks elsewhere, so that a situation would arise in which each oil-consuming state would have borrowed from the oil producers enough funds (not cash, but rather a promise ultimately to pay) to cover the extra cost of its imported oil.

The experience of 1974 showed that the practical difficulties are much greater than the theories would suggest. First, the developing countries of the world, on whom the increased price of oil bears more severely even than in the industrialized countries of Western Europe, are not a part of the international banking machine and are excluded from recycling. Secondly, even within the international banking community of nations prepared to exchange each others' currencies, some countries are less creditworthy than others. Through much of 1974, the government of Italy was able to cover its need of credit only by essentially political agreements with other industrialized countries, West Germany in particular. Britain, equally vulnerable economically, was for the most of 1974 the fortunate recipient of oil surpluses. Thirdly, the amount of surplus credit in the banking system has been so large that, with the best will in the world, the banks have been unable to digest it easily – to the extent that the surplus revenues of the oil producers are placed on short-term deposits and are able to be withdrawn (in principle at least) on a few months' notice. The commercial banks are understandably (and in many cases legally) unable to promise that they can safely pay interest on these deposits if the only way of doing so is to re-lend the surplus funds for a longer period to governments or industrial users.

By the second half of 1974, it had become clear that the amount of the oil producers' unused credit slopping around in the international banking system was too much for comfort. At a meeting of international bankers at Williamsburg, Virginia, in August 1974, Mr David Rockefeller, president of the Chase Manhattan Bank, said that the international banking system could not handle the inflow of credit from the oil producers for much more than a year, whence there has arisen a variety of schemes for turning oil producers' short-term deposits into more permanent financial investments. The United States government has canvassed widely schemes for persuading oil producers to invest in medium-term government securities issued by the oil consumers, if necessary guaranteed against inflation. Mr Denis Healey, Chancellor of the Exchequer of the United Kingdom, urged that the International Monetary Fund (IMF), at its meeting in October 1974, should set up an international fund that would save essentially the same purpose of converting the oil producers' short-term lending into a financial asset that would be more dependable. The United States favoured an international fund more directly under the control of the oil consumers.

In reality, the problem is at once simple and political, but not financial. To the extent that the oil producers are earning more than they can spend, they are building up an obligation by the oil consumers to pay for crude oil currently consumed in goods and services at some future date. In round numbers, the annual accumulation of credit (or debt by the oil consumers) is 3 per cent of the GNP of the industrialized world. By 1980, the accumulated credit may be as much as a quarter of the annual production of the West. Politically, circumstances in which the oil producers could turn their credit with the oil consumers into goods and services simply by declaring that the time had come when it suited them to convert promises to pay into motor cars, wheat or some other commodity, are inconceivable. Even in financially old-fashioned countries such as the United States, the time has long since passed when the owner of a dollar bill could demand that the US Treasury give him the equivalent in silver – all he can expect is another dollar bill in exchange. In the same way, the producers must recognize that when the time comes for them to turn their accumulating credits partly into goods and services, they will have to buy them on the international markets.

What this implies is that neither the lenders (producers) nor the borrowers (consumers) can at this stage know how the accumulating obligations will be requited. The situation is unsettling for both partners. Since 1974, it is the borrowers that have made most of their difficulties, drawing attention to the problems of recycling surplus funds, seeking in individual ways to moderate the accumulation of their borrowing from the oil producers by using less oil or even less of all kinds of commodities, raw materials and manufactured goods. But the problem also disturbs, or should disturb, the oil producers. One cannot, after all, eat dollar bills nor be sure that a sackful of dollar bills can be converted, at a snap of the fingers, into a fighter bomber, a refrigeration plant or a lathe. If the oil consumers should at some time in the future be either unable or unwilling to supply such goods and services, the most that the oil producers can expect to get in exchange for their accumulated currency reserves will be a numerically equivalent quantity of paper money, devalued by whatever fraction the potential suppliers of goods and services find appropriate. In other words, the oil producers' accumulating fortunes are hostages to the governments of oil-consuming states, at least so long as the oil producers let their funds lie fallow, merely earning interest at rates which may turn out not to keep pace with inflation.

Logic would dictate that the lenders and the borrowers should come to an agreement about the way in which the accumulating debt would ultimately be liquidated. The only practical solution is that short-term loans by the oil producers to the West should be converted into long-term credits. So long as the question remains undecided, the character of the unrequited credit is potentially a much more serious cause of financial instability and political tension than the increased price of oil itself.

The delay in facing this problem was one of the most conspicuous failures of the oil consumers in the months following January 1974, and it is part of the general failure of the oil consumers, and especially of the industrialized states, to agree among themselves about their common interest. The initial reaction of governments throughout the West was to embark on economic and fiscal policies devised to reduce by 10 per cent or so the rate at which their obligations to the oil consumers were accumulating. By April 1974, the governments of France, Germany and the United States had taken steps to reduce their deficits on their balance of payments, counting the oil deficit as a real obligation. The government of Italy followed with the fiscal device of import deposits intended to restrict the volume of imports.

So long as the OPEC surpluses are unrequited by supplies of goods and services, however, oil consumers can reduce the rate at which they accumulate obligations only by shifting some of their burden to others. By the time of the IMF meeting in Washington in October 1974, the lesson had sunk in, and there was general support for a proposal that at least a part of the OPEC surplus should be placed directly with the IMF, not the commercial banks. The sticking point for many oil consumers, the United States especially, was whether the surplus oil producers should at the same time acquire the influence in the management of the IMF, and thus of the world's monetary system, to which their supply of credit would normally entitle them. But for as long as the surpluses exist, there is no alternative to some such scheme.

The experience of 1974 provided other discouraging signs of the failure of the oil consumers to act in their common interest. Indeed, the immediate outcome was an unseemly scramble for oil. Governments like the British were resentful of the way in which the international oil companies were able partly to supply the needs of countries such as the Netherlands from sources not covered by the embargo, forgetful of the way in which this kind of flexibility had helped them during previous interruptions of the oil supply. Two British ministers made a humiliating journey to call on the Shah of Iran in Switzerland at the turn of the year, and returned with a promise of five million tons of Iranian crude oil at a price not substantially different from the tax-paid cost of oil to the companies operating in the Persian Gulf. The French government followed a different course, the first objective of which was to sell military and engineering equipment to Saudi Arabia and Iran in quantities that might be expected to offset a substantial part of the French oil deficit. By the end of 1974, Iran had agreed to buy $6000 million worth of French industrial equipment by 1980, and in December it became known that Iran would also invest in a plant being built in France (with the partnership of Belgium, Italy and Spain) to produce enriched uranium.

The first response of the United States to the events of 1973 was to reinforce the theme running through President Nixon's declarations on energy during 1973 that the United States should set out once again to be self-sufficient in energy production. 'Project Independence' was the slogan. This was quickly followed by the view that the United States should take the lead in organizing a common front by the oil consumers, an ambition of the Secretary of State, Dr Henry Kissinger, initially frustrated by the unwillingness of the French government to agree to most of the proposals for common action put forward at the international conference in Washington in February 1974.

French objections to taking part are important and deserve respect. The essence of the case is that the national interest of France will be best served by ensuring a steady supply of energy, that for the next decade OPEC oil is essential and that the best way of securing this is by the forging of links with oil producers that go beyond the largely political arrangements devised in the past. The agreements reached with Iran during 1974, if successfully carried out – and it remains to be seen whether French industrial companies can supply the equipment and the technical assistance promised on their behalf –

will probably ensure a continuing share of Iranian oil for France; but at a price.

As an extension of this argument, the French government for most of 1974 consistently opposed attempts by other oil consumers to develop common policies that might seem to the oil producers hostile. This is analogous to the United States view in the early part of 1974 that formal negotiations between oil producers and oil consumers would institutionalize OPEC and give tacit consent to oil prices and terms of supply which can be shown to be unjust.

Fortunately, by the end of 1974, there were signs that these differences of view were being eroded, and that the foundations of an institutional framework for dealing with the genuinely international aspects of the oil problem were appearing. The international oil conference in Washington in February 1974 led to the formation of a Petroleum Policy Consultative Group, transformed at the end of 1974 into an International Energy Agency. Although France is not a member, it has been agreed that in due course there should be a conference of oil producers and oil consumers to discuss an agenda not originally defined.

What should be the common objectives of the oil consumers? In what way can the oil consumers acting in concert bring more rationality into the international supply of energy than could be done by individual nations acting separately? How can negotiations with the oil producers help?

The first need is to regulate the international monetary problems created by the increase of the price of oil. To the extent that some increase of the price of oil above the prices obtaining in the 1960s is just (and certainly inescapable) there is bound in the long run (when the oil producers can spend what they are now earning) to be a substantial transfer of economic resources from the producers to the consumers. No amount of international negotiation can make that simple truth go away. But it is in the interests of both parties that oil surplus investments should go to the oil consumers on terms which do not consititue a threat to international monetary stability.

There is nothing in the history of the past few years to suggest that OPEC members are anxious to encourage financial instability in the West, however much signs of it may amuse them. They have too much to lose. But this does not imply that persuading them to put their surplus funds on long-term deposits is simply a matter of persuading OPEC members to be reasonable. Understandably, they will ask how their funds will be protected against inflation and devaluation. The oil consumers themselves will be uneasy at committing themselves to pay interest for several years at the high prevailing rates. But the fact that both sides must compromise at least suggests that meaningful negotiations should be possible.

The problem of recycling surplus funds, essentially an internal matter for the oil consumers, is not a mere technicality but a political issue of great potential difficulty. No system of recycling surplus funds can function independently of an appraisal by the financially strong oil consumers of the way in which their weaker brethren, to whom the surplus funds would not normally flow, conduct their internal affairs. In the past few years, both the British and the Italian governments have experienced some of the discomforts of having their internal policies overseen by the IMF. In the long run, such inquisitions are likely to be inseparable from any attempt to even out the flow of OPEC surpluses. The increase of oil prices may ultimately have the effect of bringing the oil consumers closer together, but in the most painful fashion imaginable – the relationship between reluctant creditors and resentful debtors.

A further issue that needs to be settled is the justice of the increased price of oil and the terms on which OPEC members have acquired participation in their oil production from their concessionaires. As has been seen, the present price of oil is unreasonably high in two senses: it is greater than the cost of exploiting alternative sources of energy in the decades ahead, and it will force the oil consumers to invest in capital developments on a larger scale than would otherwise be necessary – tantamount to an uneconomic use of the world's resources. To be sure, OPEC members are sovereign states, and are entitled to price themselves out of the market in the long run. But there is every reason, in everybody's interest, why the oil consumers should set out to put OPEC's new prices to the test. Since alternative sources of energy will not be available in substantial amounts before the 1980s, the only feasible way in which a serious test could be made of OPEC's intentions would be a concerted if temporary reduction of oil consumption by the oil-consuming nations. And whatever the legal rights of OPEC countries to fix their own prices, there is plenty of room for argument about the terms on which participation has been secured.

To establish a common front on all these issues will be exceedingly difficult, especially because effective concerted policies will require oil consumers to sacrifice some of their sovereign independence if their common policies are to succeed. The treaty establishing the International Energy Agency requires signatories to share their energy resources at times of scarcity, but it remains to be seen whether the British government will accept that this applies to oil from the North Sea; or whether the United States Congress will agree that Texan oil should be exported to meet the needs of Europe or Japan.

In all the circumstances, it is unlikely that the dialogue that must eventually come between producers and consumers will be easy or quick. But it is urgent. And so are the steps which the oil consumers must take to provide those alternative sources of energy that would anyway have been necessary before the end of the century, but which are now forced on us prematurely.

7
Wasting and wanting

The rapid increase in the price of oil has powerfully stimulated an intermittently recurring theme in the affairs of industrialized societies: the notion that it should be possible to solve a great many problems not by exploiting new kinds of fuel but by reducing energy consumption. 'Energy conservation' is the slogan, and it is beyond dispute that a calorie not used is a calorie that does not have to be produced by burning fossil fuels or in some other way. In present circumstances, energy conservation has often come to seem a virtuous objective in itself. Unfortunately, energy conservation is not a form of self-denial, like giving up alcohol. Usually, it costs capital and resources. What might energy conservation accomplish in present circumstances? Is waste a partial explanation of our troubles?

Except in proverbs and homilies, waste is not an absolute but a complicated amalgam of economic considerations. Energy consumption is never an end in itself, but a means to an end. People purchase electricity because they want to be warm, or value the convenience of a refrigerator. People buy gasoline because they want to be able to travel in automobiles from one place to another. Manufacturers burn coal or oil not for atavistic reasons, but because they wish to make and market products. And in all these circumstances, the purchase of energy is only one of several purchases that must be made before the end result can be obtained.

The only meaningful use of the concept of waste is therefore one based on the total value of the economic resources needed to accomplish some specific goal. In many circumstances, relatively accurate economic arithmetic is possible. For example, if the objective is to provide motive power for industrial operations, as for turning lathes in engineering factories, it is plainly an economy of resources that electricity should be manufactured at a generating plant and distributed to those who use it rather than that individual manufacturers should generate their own electricity. One 1000-megawatt generating station and the associated transmission lines would cost much less than ten thousand 100-kilowatt generating plants, a simple reflection of the fact that a large plant can be made with smaller amounts of raw materials and less labour than can a multitude of small generating plants. But this is also a situation in which the amount of energy consumed for each unit of electricity produced is smaller in the larger plant, for however much heat may be discharged into the environment of large electricity generating plants, the efficiency of the smaller units would almost certainly be far less. And this calculation, of course, takes no account of the convenience and money savings to industrial consumers of electricity because they do not have to make arrangements for keeping their machines at work if their own generating plant should break down. Hence 'waste heat' is not waste at all, but an accompaniment of an economically efficient activity.

The economics of the much maligned automobile are far harder to assess, chiefly because it is difficult to quantify the benefits that automobile owners derive. It has been approximately calculated that travelling 1000 miles by motor car in the United States entails the use (taking the city streets with the interstate highways) of about the energy content of 1.4 barrels of crude oil. If energy consumption were all that mattered, walking would be much more economical, consuming only a twentieth of a barrel of oil, supplied as food and expended as mechanical energy. Not even the most zealous of those who consider energy conservation to be an end in itself would infer that they should walk 1000 miles. For one thing, they would find that their total consumption of energy would be three times that expended merely in walking, for they would have to breathe and keep warm while resting. In reality, of course, most people simply cannot afford the time to walk 1000 miles. This absurd comparison is therefore not merely a striking illustration of the discovery of the 1920s that the automobile is a liberating device, but also a proof that it is a source of rational economic benefit.

In the United States in the early 1970s, transport accounted for 25 per cent of all energy consumption, of which automobiles accounted for just over a half. In Western Europe, the proportion of total energy consumption in transport as a whole was more like 10 or 15 per cent. In Britain, road transport consumed nearly 12 per cent of all energy, and motor cars 60 per cent of that. All things considered, it is hard to find evidence that the value of private transport to those who own motor cars is altogether different from the price they are prepared to pay for it. In any case, the cost of private road transport is largely the cost of paying for the car, not for the petrol it consumes, even at the prices now posted at the pumps. In 1972, in Britain, the cost of private road transport was 10 per cent of all personal expenditure, with petrol accounting for less than half of this.

The undisputed facts that most car journeys are short journeys, that many of them could be accomplished by public transport and that cars create traffic congestion are not a proof of the malevolence of drivers or of the inutility of the automobile. They are, if anything, the opposite – that those who travel by car consider it worth incurring substantial inconvenience and cost to do so. And it goes without saying that, especially in the United States, the geographical patterns of life that have developed in the past few years outside the major cities would be insupportable without automobiles, and that they could only be quickly re-created in another mould by the investment of very large sums of money, even larger than whatever reductions of any country's import bill would be occasioned by a reduction of gasoline or petrol consumption. In short, the fact that in spite of everything motor car travel remains conspicuous is in itself a proof that, from the point of view of individuals, car ownership is a good use of economic resources.

In the circumstances, the question that should be asked by opponents of road travel and those who see road travel as a large potential saving of energy is not how the use of motor cars in particular and of other froms of road transport as well can be curtailed by government regulation, but whether those who use road transport are required to pay the full social costs involved.

The subject is enormously complicated. In Western Europe, taxes on road transport and on petrol have been for several years greater than the sums of money spent on building new roads and operating the licensing machinery for road vehicles, but no community has yet taken steps to saddle vehicle operators with the social costs of traffic congestion. In the United States, federal taxes on gasoline have for several years been used to finance the construction of inter-state highways, but state sales taxes on gasoline have not been sufficient to cover the cost of social investments on behalf of road transport. For several decades it is likely that the cost of gasoline in the United States has been less than it should have been, which is no doubt part of the reason why gasoline consumption is higher in the United States than elsewhere (and also, in the long term, why the density of population even near major cities is lower than in Europe), and why it has been increasing rapidly for several decades.

Indeed, the disparity between the cost to a car owner of driving, even in urban areas, and the true cost of his activities to the community is so great that it is no wonder that economists, especially in the United States, are continually marvelling at what they describe as the inelasticity of the demand for gasoline – small increases of the price of gasoline have in the past seemed not to have affected the growth of demand for gasoline significantly, large increases of price have brought only small reductions of demand. The truth is that the market does not function as the textbooks say it should, because the costs of operating motor vehicles do not accurately reflect the social and other costs, among which might be included the high cost of producing gasoline in oil refineries.

Motor transport, then, is a field in which energy economies may most efficiently be won not by overriding the market mechanism by further regulations on the use of motor vehicles but by making sure that the market mechanism has a chance to function efficiently. In the United States, one of the absurdities of the past few years has been the attempt by Congress, the administration and state governments to legislate for cleaner air, especially in urban areas, by requiring that motor vehicles should be fitted with devices for controlling the emission of carbon monoxide, unburned hydrocarbons and nitrogen oxides, with a consequent increase of energy consumption variously estimated at between 10 and 15 per cent, when they might have achieved the same result, and a measure of energy economy as well, by devising some means of making sure that those, but only those, who added to urban air pollution were required to meet the costs.

Against this background, the schemes devised in recent years for saving energy on motor transport are unrealistic. Thus the Ford Foundation's Energy Policy Project calculates that it might be possible to reduce anticipated energy consumption at the end of the century (in A.D. 2000) by some 40 per cent by using such devices as an increase of the average fuel consumption of cars to twenty-five miles per gallon (1 US gallon=0.8 Imperial gallon), the transfer of short-haul air traffic to railways as well as the transfer of some inter-city freight traffic. But the project has nothing useful to say about the ways in which these changes would come about, nor about the costs that would be entailed in providing these alternative methods of travel. And in the calculation of what would be entailed in its scenario of Zero Energy Growth, in which United States consumption of energy as a whole does not increase after 1985, the project says that the amount of energy needed to drive automobiles would be markedly reduced because people 'to save the time wasted in commuting . . . would live closer to work or closer to schools and shopping areas' without referring to the cost of rehousing on such a scale and without explaining how this vast change in people's working habits would be brought about, except by taxation policies to discourage home ownership in the suburbs.

For road transport, and particularly individual road transport, the moral is clear. Those who look to these activities for potentially large reductions in the demand for energy are probably right to do so, but have not yet acknowledged the need first to ensure that the market that should efficiently allocate economic resources is given a chance to work; and, secondly, that the period of time in which substantial changes are likely to come about may be quite long. It is unlikely that substantial economies in gasoline consumption can occur in periods shorter than the interval of time between the replacement of motor cars. In the special circumstances of the United States, where gasoline consumption accounts for 25 per cent of all energy consumption, there is, however, a case for using excise taxes, at steadily increasing rates, as a way of exploring the elasticity of demand.

Similar considerations apply in other fields where large amounts of energy are at present used inefficiently. Domestic heating is a conspicuous example. In the last resort, of course, all the energy used to heat a dwelling escapes to the environment. The practical question is how quickly it does so. Especially because energy has traditionally been, and is still, sold at unrealistically low prices, house owners have not a sufficient incentive to insulate their buildings to the standards that would now be economic. In 1973, in the United States, the Federal Housing Agency introduced new standards that will in future apply to all private dwellings qualifying for federally assisted mortgages (in which space heating accounts for 11 per cent of the total consumption of energy in the United States), and are likely to be widely applied. However, this is again a field in which changes will come slowly – the cost of insulating a house that already exists is necessarily greater than the cost of insulation in a new dwelling, which means that it may be a sensible use of economic resources to go on burning larger amounts of fuel instead of footing the bill of higher insulation costs.

In Britain, a working party of the National Economic Development Office estimated in May 1974 that improved insulation of British houses would often be used not to reduce energy consumption but to provide more comfort. Even so, the working party estimated that the overall result would be a 10 per cent reduction of the demand for energy for space heating, itself the equivalent of 56 million tons of oil a year if allowance is made for the inefficiency of power stations and domestic furnaces. What this implies is that each of the country's 19 million dwellings might, on average, reduce its gross energy consumption by the equivalent of two barrels of oil a year, which is the equivalent of an average cost of £17 a year. With interest rates at 16 per cent or so, it would plainly be uneconomic for the average individual to invest in extra insulation unless he could be sure of a 10 per cent saving of energy consumption for an initial cost of less than just over £60, much less than the estimated cost of insulating new houses to an acceptable standard. To rehearse this simple and stark arithmetic does not imply that all domestic thermal insulation would be valueless, but merely that it would be, in present circumstances, an enormous waste of economic resources to insulate all houses to a high standard. Much better to import oil, even at OPEC prices.

In Western Europe, as in the United States, there is therefore not so much a need to decree measures of energy conservation as to create the conditions in which they will be adopted for economic reasons. In Britain, during 1974, the government seemed to be bent on doing the opposite. Throughout the year, for example, the nationalized industries responsible for coal, gas and electricity were running at deficits corresponding to something like 15 per cent of their total sales of energy. In October 1974, the Central Electricity Generating Board estimated that its deficit for the then current year would amount to £250 million. In other words, for all the talk of 'energy crisis' and the British government's evident concern for the financial problems occasioned by the increase in the price of oil, consumers of energy were still being protected from the full cost of what they were using, with the result that consumption was higher than it would otherwise have been, and that the industries responsible for meeting demand were unable adequately to provide the capital equipment needed to ensure future supplies. That the external financial problems of countries such as Britain were, by these devices of price control, exaggerated was another consequence.

In reality, even the most dirigist of governments must acknowledge that if they interfere with the workings of the economic system to reduce prices for scarce commodities, they are bound to increase demand and reduce supply, thus accentuating the scarcities with which they have set out to deal. The external financial problems of the oil consumers being as serious as they are, there is no justifiable cause for selling energy more cheaply than it can be produced. With a realistic pricing policy, however, and a good supply of patience, governments would find that the reductions of energy demand on which many of them have set their hearts would ultimately materialize.

In the meantime, there are many steps that governments could take to encourage long-term changes in the patterns of energy consumption, leading, in due course, to substantial economies. One common characteristic of situations in which energy consumption might be reduced – road transport, dwellings, commercial buildings and industrial processes – is that energy is needed for the proper functioning of large items of capital equipment. It is unreasonable to expect a manufacturer to replace an expensive machine tool, the cost of which may be a hundred or a thousand times the cost of energy it will consume during its lifetime, simply because a machine that uses energy more efficiently has appeared on the market. This would be a waste not merely of the manufacturers' but also of the community's economic resources. There are, however, many circumstances in which the renewal of inefficient plant or equipment would be economic, but where replacements are not made because those who make the decisions either do not have access to the capital needed for renewal or are simply ignorant of the possible economic advantages. And there are many circumstances, particularly in industry, where inertia, intellectual or financial, inhibits the development of new processes that would be economically advantageous and that would also save energy. The reclamation of energy from domestic refuse (a potentially useful if not very substantial source of energy) and the introduction of new techniques for extracting aluminium from its ores are two conspicuous examples.

It follows that governments need energetic organizations to ensure that those who consume energy are fully aware of the economic costs of their established practices. After a decade in which consumerism has made rapid progress in most industrialized societies, it is anomalous that few governments consider it part of their responsibility to remind those who buy houses or motor cars of the trade-off between the purchase of a badly insulated house or a needlessly heavy motor car and the fuel bills they will have to pay in the years ahead. (In the

United States, the Environmental Protection Agency has broken new ground by making an assessment of the fuel consumption of different makes of automobile, and during 1974 purchasers of smaller cars in the United States were increasing rapidly, though not necessarily as a direct result.)

It should not be beyond the wit of governments also to devise means of making sure that energy consumers are not so inhibited as at present by shortages of capital from making economically sensible decisions about the purchase of energy-consuming equipment. Indeed, it is in many ways surprising that entrepreneurs have not already recognized this to be a fruitful and potentially profitable field of activity. A person who buys a motor car, for example, would probably find it advantageous to buy a package consisting of the vehicle and the supply of petrol needed to keep it on the road. People who buy houses might similarly find it both convenient and an incentive to the efficient management of their economic resources if their mortgage payments included nor merely the amortization of the cost of the house but the purchase of energy. In industry, the problems of capital investment in economically efficient plant which makes the best use of energy are theoretically much simpler, for industrialists are supposed to be hardheaded, calculating fellows. But everyone will acknowledge that the shortages of capital for industrial investment, which have themselves been caused by the financial climate occasioned by the increased price of oil, are extra inhibitions to the efficient use of such industrial investments as exist. For all the theoretical calculations that have been made in recent years of the amount of energy that might be saved in industry, it is likely to take several years before the necessary capital investments are made.

The experience of the past few decades has shown, clearly enough, first that while improvements of energy efficiency are inherently a part of industrial development, they are attainable only over decades. The electricity utilities in all industrialized countries are a striking illustration. In the United States, for example, the amount of energy needed on the average to generate a kilowatt hour of electricity was reduced by just over 35 per cent between 1948 and 1968. The technical explanation is simple. More efficient generating sets were brought into service, bulk transmission lines were introduced so that electricity utilities could keep the most efficient generating plants more consistently at work and devices for storing electricity (by pumping water uphill into storage reservoirs) were used to the same effect.

In the steel industry, the increase of the efficiency of manufacture has been even more dramatic. In Britain, for example, the energy needed to produce a ton of steel decreased by 24 per cent between 1962 and 1972, chiefly by the introduction of the basic oxygen process. The effect of this new process on the energy of individual steel-making plants has been almost revolutionary. At the Lackawanna plant of the Bethlehem Steel Corporation in the United States, for example, the energy needed for each ton of steel decreased from the equivalent of 22 gallons (US) of oil to 6.4 gallons of oil per ton. (The steel-making process excludes the production of pig iron, in which energy consumption per ton is much higher.) It is, of course, inevitable that steel-making plants do not march perfectly in step with each other, introducing the same technical improvements simultaneously, which is why the effects of technical innovations on energy consumption are bound to be spread over a substantial period. What does emerge is that especially in the industries where the consumption of energy is an important factor in the cost of production, improvements of efficiency are usually steady and, cumulatively, of great importance. This, after all, is what marked the development of the steam engine a century ago. The practical task for governments concerned with energy conservation is to ensure that the forces which make the major energy-consuming industries change are also sensed elsewhere in industry.

None of this can excuse the failure of governments adequately to finance the kind of research and development that could potentially bring both economies of energy and more efficient use of economic resources. While, in 1974, the United States did embark on a programme of research and development in the efficient use of energy which was at once imaginative and ambitious, elsewhere, the scale on which governments in the West are supporting research and development is hardly commensurate with the scale of the financial problems that efficient energy utilization might alleviate.

Technically, the objectives are almost self-evident. One of the most important is to develop the means of storing energy in bulk, for this would make it possible for electricity utilities to increase the proportion of electricity generated by the most efficient plant. In the past decade, utilities throughout the world have sought to store energy from times of the day when demand is low to times of peak demand, but the potential of pumped storage is limited. Now, however, there are possibilities of using massive and rapidly spinning fly-wheels for the same tasks. Improvements of the efficiency with which different fuels are burned are possible and potentially important. So, too, are devices by which the efficiency of electrical generation might be substantially improved – the coupling together of gas turbines and more familiar turbo-generators, the use of electrically conducting streams of hot gas to generate electricity in machines without moving parts (magnetohydrodynamics) and the possible use of hydrogen as a fuel. The possibility that new types of vehicles for surface transport might be in the broadest sense economically advantageous is real enough, although the electrically driven motor car is technically more difficult than is often supposed.

It is only fair to acknowledge that the management of research and development of this kind presents governments with awkward and unfamiliar problems. In the nature of things, most improvements in energy efficiency come about in humdrum ways, by the step-by-step improvement of the design of machinery. The organizations responsible for design, usually the larger plant-manufacturing companies, are those whose interests and skills best coincide towards such objectives. What governments can undertake effectively is the development of more radical innovations. The difficulty is that long-term and usually expensive research programmes can fail, if not for technical then for economic reasons. Although government support of long-term research and development programmes is now urgently needed to lend more flexibility to the pattern of energy supply, there is a danger that too much pre-occupation with the technical basis of energy conservation may divert governments from their equally necessary support of more widely directed long-term research and development whose benefits might include increases of economic resources that are, in the last resort, alternatives to energy conservation or the provision of alternative supplies of energy.

For what it is worth, the creation in the United States of the Energy Research and Development Agency out of the framework of the old Atomic Energy Commission promises to be a workmanlike way of implementing the research and development programme devised in December 1973 at an estimated cost of $10 000 million over ten years. The scope of the organization is large enough to allow the potential benefits of different kinds of energy innovations, conservation and new means of producing energy to be judged alongside each other without being so large as to distort the balance of research and development in the United States as a whole. In Europe, progress has been slower.

The implications are clear. True energy conservation, the more efficient use of energy in the support of what an industrial society does at present, is bound to be a slow process and may use economic resources on a substantial scale. The role of governments is chiefly to ensure that energy is sold at realistic prices, that the advice people need to make rational decisions about energy consumption is freely available and that the lines of research that will yield energy-saving innovations are properly funded and energetically pursued. That done, they must sit back and wait.

It is, however, possible to reduce a community's total consumption of energy by more direct intervention, and the past few years have seen a number of essays of this kind, some of them intelligent, some less so. Reduced speed limits for motor vehicles, now common in oil-consuming countries, are simple and enforceable but are unlikely to bring large savings. Regulations specifying the maximum temperature of office and domestic buildings, introduced in the United States at the end of 1973 and in Britain a year later, are potentially more valuable but less easily enforced. A mixture of excise taxes and import duties is a potentially useful means of making consumers appreciate that it may be in the national interest to reduce consumption, but logically they should be applied not merely to sensitive materials such as petroleum but to other sources of energy as well.

None of these devices could, however, be expected to bring

about the short sharp reduction of energy consumption that would be necessary to put the new OPEC prices to the test. If the oil-consuming states could persuade themselves – as they should – that such a test is necessary, their most practical course will be to ration uses of energy that do not contribute directly to industrial and agricultural production. Much of the interest of the period ahead will be to see whether democratically elected governments can bring themselves, in concert, to adopt such necessarily unpopular measures, however temporarily.

8
Where else to turn?

Energy conservation, as has been seen, is not an immediate solution to the problems caused by the increase of oil prices. In the long run, it is likely only to moderate the speed with which energy consumption increases. So what else can be done to find alternative streams of energy? In the past few years, there have been several quite separate threads in this discussion. First, it has been argued, better use could be made of the fossil fuels, not merely of coal but also of petroleum still to be discovered and other sources of hydrocarbons such as oil shale and tar sands. Secondly, the world's energy resources could be augmented by the application of entirely new techniques – power from the sun, from the sea or even from the interior of the earth. Finally, there is nuclear power (see Chapter 9), which happens to be a way in which the energy resources of the world could be increased substantially in the near future, but which is nevertheless something of a mixed blessing, chiefly because of the safety problems involved.

The return to fossil fuels will not be a solution to all our problems, but only a temporary expedient. Jevons identified more than a century ago the essential characteristic of fossil fuels – that the cost of extracting them is bound rapidly to increase, not merely as the most prolific deposits are worked out, but also as the prosperity they create drives up the cost of labour. It is true that in the United States there are huge deposits of coal quite near the surface in parts of the country – in Wyoming, for example – which were neglected by the early and rapid spread of the Industrial Revolution. There, still, cheap coal is to be had, provided the drag lines can be built in time. The chances are, in fact, quite high that the output of coal in the United States will in the years ahead increase substantially above the 660 million tons of coal a year (equivalent to some 450 million tons of oil) at which output has recently settled. But it is hardly conceivable that the output of coal could increase by a half. The chance that it may double by 1985, one essential feature of Project Independence, is negligible.

Britain, as one of the founders of the coal industry, provides a salutary illustration of the outstanding problems. In June 1974, the British government agreed with the National Coal Board, the nationalized coal-mining industry, that £600 million ($1440 million) should be provided out of public revenues to sustain the capital investment programme in the remainder of the 1970s. Two thirds of this sum will be spent on the replacement of existing mine capacity (which is another way of saying that the National Coal Board has not, for some years, been required to finance its own renewal out of current income). The extra coal-mining capacity is reckoned, somewhat optimistically, to be between 10 million and 15 million tons a year or, on the most generous estimate, the equivalent of 10 million tons of oil a year or 200 000 barrels of oil a day. In short, the British government is proposing to increase coal-mining capacity by investing something like the equivalent of $2,500 for each daily barrel of equivalent oil production, slightly cheaper than the cost of winning oil from the North Sea but not much cheaper, and much more expensive than the cost of producing power from nuclear power stations.

British experience is not directly applicable in North America (though it is relevant enough to the rest of Europe for 1974 to have passed without more than token signs of encouragement for the conventional coal industry in Germany and France). In the United States, however, there is much more scope for the profitable expansion of the coal industry. Not merely is coal extraction more economic, but the coal – in itself an inconvenient fuel – can more easily be converted into other forms. The plan for energy research and development in the period 1975–9 submitted to the United States Administration by Dr Dixie Lee Ray, chairman of what was then the Atomic Energy Commission, advocated that a fifth of a total of $10 000 million should be spent on research and development in coal, on projects that included the improvement of mining methods, the more efficient burning of coal and the development of techniques for the conversion of coal into gas and liquid hydrocarbons. The costs of gas comparable with natural gas in thermal content and chemical composition is reckoned to be something like $1.00 for a million BThU, which corresponds to just over $5.00 for the energy content of a barrel of crude oil. The cost of turning coal into liquid hydrocarbons – by an elaboration of the process used in Germany in the Second World War – is more uncertain, though some studies carried out in the United States as a basis for the present plans for energy research and development suggest that production costs might work out at roughly the same as those of synthetic gas. The overriding question is whether these processes can be developed and exploited quickly enough significantly to affect the United States energy market by the 1980s.

The speed with which the scale of coal utilization can be increased is also limited by the technical problems that will have to be solved before large numbers of synthetic gas and petroleum plants can be built, as well as by the variety of environmental regulations that now restrict the way in which coal can be burned. By 1974, American electricity utilities had spent between $300 and $400 million on equipment for removing sulphur dioxide from the flue gases from electricity power station boilers, but there is still no assurance that the years ahead will see the development of economical method of meeting the environmental standards now imposed on sulphur emissions from plants which burn raw coal. Organizations such as the Tennessee Valley Authority have argued that their operations should be regulated not by rules specifying how much (or how little) sulphur there should be in their flue gases, as by regulations specifying that the atmosphere in the neighbourhood of their power plants should not be polluted above specified limits, which would make it possible for power plants in rural areas to make direct use of coal otherwise unusable in its raw form. There is unfortunately little chance that any such sane proposals will emerge from Congress without difficulty and delay.

Yet the issue is important, not only for the economics of power production in the United States, but also for the strategic relationship between the oil consumers and the oil producers. For if, in the next ten years or so, the United States should be able to produce a substantial part of its energy at a cost not very different from $5.00 a barrel, and if there is a prospect of increasing this supply substantially, the bargaining position of the oil consumers in their relationships with OPEC will be greatly strengthened.

The possibility that other kinds of fossil fuels might be exploited as sources of energy, and, in particular, that the oil shales and tar sands of North America may become important contributors to the American energy economy in the years ahead, is more problematical, for the reasons already given in Chapter 3. That the oil shales of Utah, Wyoming and Colorado are potentially enormous reserves of energy is beyond dispute. Whether they can actually be made to yield more than a small fraction of their potential is another matter. The Department of the Interior has sensibly agreed that no further tracts of the federal lands will be let out for oil shale workings until 1985, and it is unlikely that the oil shales of the western states will make any large contribution to the production of synthetic petroleum unless, unexpectedly, it does turn out to be possible to extract petroleum from the oil shales without physically mining the rock. Schemes for extracting the hydrocarbons in which the shale oil is effectively distilled from the underground strata may yet prove practicable.

In the long run, the amount of shale oil produced in the United States will be determined by the as yet unknown economic balance between shale oil and coal, the environmental problems of exploiting oil shales on a large scale (among which shortages of water in the Colorado River Basin is likely to be the most serious obstacle to expansion beyond the scale of production now planned) and the likely balance of the world's energy supplies beyond 1985. Estimates of the cost of producing syncrude from oil shales current during 1974 range from the equivalent of $4.00 to $7.00 a barrel, figures likely to be increased rather than decreased in the years ahead as the difficulties of working underground mines (in which some of the seams of oil shale will be more than 100 feet thick) and complex chemical engineering plants in some of the most remote regions of the United States are translated into fact.

The prospects for increasing the yield of other potential sources of fossil hydrocarbons in the years ahead turn on the likely availability of crude oil and natural gas from regions in the world not so far exploited. In spite of the excitement generated by recent discoveries of petroleum reservoirs in southern Mexico, and which may in the years ahead return Mexico to the list of the world's oil exporters, the chances are that future discoveries of a size sufficient to affect the pattern of the international market will be made in offshore oil fields – the China Sea, off the east coasts of South and even North America and the rim of the Arctic basin (where the North Slope of Alaska has so far been an accurate guide to what may be found elsewhere). In the past few years, enthusiasm for the potential of these largely inaccessible parts of the world has too frequently been uninformed by a proper appreciation of the costs.

Although, after the discoveries of oil on the North Slope, there were some in the United States who considered that the availability of petroleum was no longer a problem, the truth is that the North Slope is unlikely to yield much more than 150 million tons of crude oil a year at a cost in excess of $5.00 a barrel, already inflated by a delay of more than three years in its exploitation. The influence of North Slope oil on the international market is thus likely to depend on the degree to which the recent growth of petroleum consumption in the United States persists. If, as is more than likely, the pace of growth of consumption is now moderated, and especially if petroleum consumption remains static, Alaska could account for between a fifth and a quarter of total consumption and thus would be a powerful influence to moderate oil prices in the region of $5.00 to $6.00 in North America and, say, $7.00 in Europe. If, on the other hand, the growth of demand for petroleum in North America is resumed at anything like the pace characteristic of the 1960s, Alaskan oil will have very little influence on the price to OPEC.

In Europe, the oil fields of the North Sea are plainly of

great regional importance, and if the pace of growth of oil consumption now moderates, they may account for a fifth or more of European consumption by 1985. The snags are principally that it is not yet possible to estimate the costs of North Sea oil accurately. That from the southern oil fields may entail landed costs to the oil companies of $3.50 to $4.00 a barrel, plus royalties and taxes. But in the still incompletely explored northern tracts of the North Sea, costs could be twice as great and potential supplies thus limited by economic considerations and, ultimately, by the level at which OPEC prices eventually settle.

Offshore oil deposits elsewhere in the world are unlikely to have an important influence on the relationship between the supply and demand of oil between now and 1985, chiefly because of the time needed to determine the scale on which a newly discovered oil field should best be exploited and then to install the necessary platforms and pipelines. Although the potential of the continental shelves is probably as great as the geologists are fond of saying, and may have an important influence on the industrial development of regions such as India (where it is likely that offshore exploration now under way will yield valuable results), the extra supply of conventional petroleum between now and 1985 will consist largely of the output of the North Slope of Alaska, the North Sea, the Gulf of Mexico, the California coast and the coastal waters of Australasia. The extra production of oil may, however, amount to 500 million tons a year – by no means enough to meet the world's demand for petroleum after a decade's growth at the pace of the 1960s, but certainly enough to have a powerful influence on the international price of oil, if only the growth of petroleum consumption can somehow be constrained.

Any measures that can be taken to reduce the consumption of petroleum will have an especially important influence on the price of oil and, by extension, on the price of other forms of energy. This is why schemes for increasing oil and gas production or for producing synthetic crude from oil shale and tar sands have particular importance for the future. As always, the United States is in a position to play a unique and influential role by adjustments of its internal policies. Given the limited character of the world's petroleum reserves, it is unreasonable to expect that the industrialized world could continue to grow at the pace of recent years, that the developing countries could become industrialized and that petroleum could remain indefinitely more than a marginal source of energy (though it would still remain a crucial source of raw materials for the chemical industry). So, in the long run, nothing would be lost if the attenuation of petroleum supplies thirty or forty years from now were enhanced by accelerating the rate at which existing reserves of petroleum are exploited. The immediate strategic and monetary benefits could, however, be immense. If United States oil supplies could be increased to the point where imports of oil were unimportant – and there is even a chance that the United States could become a net exporter of oil in the first few years of the operation of the trans-Alaskan pipeline – pressure to reduce the price of oil would be enormously strengthened while the political tensions that spring from the implicit confrontation between OPEC and the industrialized countries would be reduced.

What can the United States do in practical terms? It could pursue an energetic policy of granting offshore petroleum leases; allow production from the naval petroleum reserves (possible only after a joint resolution of Congress and with the consent of the President); persuade state regulatory commissions such as the Texas Railroad Commission, at present endowed with powers to determine the maximum rate at which oil wells are exploited, that the time has come to increase these limits as far as possible, and to provide the petroleum companies with an incentive to increase production; and force on American consumers such a substantial increase of the price of natural gas that the drilling companies have an incentive not merely to find new wells but to increase production from those that exist already, a process that should in practice be easier for natural gas than for liquid petroleum.

One of the disconcerting features of United States policy during 1974 was that most if not all of these suggestions were canvassed, but were acted on only half-heartedly. By the end of the year, natural gas prices were still tightly controlled and the production of liquid petroleum had begun to increase only slowly, after falling in the first half of the year. It goes without saying that exactly the same principles could and should have been followed elsewhere in the world, even though their influence could not have been nearly as substantial as a constructive reform of American domestic policy.

In Europe, the possibility of increasing the rate of gas output from the gas fields of the southern North Sea, many of them considered too small to be exploited before the petroleum price increases of 1973, seemed not seriously to have been considered during 1974. The British government's proposals for the long-term regulation of the exploitation of the British sector of the North Sea, announced in June 1974, included not merely the acquisition of a 51 per cent share of the equity of the oil companies working there, but also that the government should have powers to regulate the rate of exploitation.

The strategic significance of the unexploited fossil fuels is two-fold. First, to the extent that supplies from non-OPEC sources can be increased in the period between now and 1985, alternative sources of energy can have a profound influence on the level of petroleum prices eventually struck between OPEC and the rest of the world. And in the longer term, the existence of the large reserves of fossil fuels still unexploited is a powerful insurance that energy in some shape or form will continue to be available for centuries to come. Precisely what relationship will eventually be established between fossil fuels and other sources of energy – nuclear power and the more exotic unconventional sources of energy – is an issue that cannot be settled in advance, for everything will depend on what the years immediately ahead suggest about the costs of producing power from sources many of which have not yet been exploited.

This is the sense in which enthusiasm for solar energy and other novel methods of producing power is irrelevant to the problems of the immediate future, to the end of the century or thereabouts. It is, of course, beyond dispute that the amount of energy from the sun that reaches the earth is very large even by comparison with the scale on which energy is at present consumed. The earth intercepts from the sun roughly 175 million million kilowatts of energy, the equivalent of more than 100 million million tons of oil a year, roughly 20 000 times the present rate of energy consumption. Tidal power is another potential source of energy, reckoned to be the equivalent of 8 million million tons of oil a year and more than 1000 times the present rate of energy consumption from fossil fuels. Geothermal energy, the energy that flows out from the centre of the earth, is also in principle another potentially vast source of energy, being the equivalent of more than ten times as much energy as is generated by the tides. It is natural enough that attention should have turned to these alternatives. The surprise is that the enthusiasts for the unconventional energy sources have so lightly assumed that the technical and economic difficulties inhibiting the large-scale use of solar, tidal and geothermal power over the past several centuries will now disappear.

In the circumstances, it is relevant that solar energy was the chief source of energy until the early decades of the nineteenth century. The water wheels of the eighteenth and earlier centuries were ultimately driven by that part of the sun's energy used to evaporate water from the surfaces of the oceans, whence it falls again as rain to swell the rivers which happen to be geographically suitable for driving water wheels. Similarly, the wood once used for domestic heating as well as for providing the charcoal needed for the reduction of metal ores was ultimately derived from the energy of the sun. And it remains the case that the food that keeps everybody alive is largely obtained from solar energy, eked out with energy obtained from fossil fuels supplied in agriculture as fertilizer and the like.

But the early decades of the nineteenth century showed clearly enough that the exploitation of fossil fuels as sources of energy for industry was of great economic and technical advantage. Even the rapid increase of the price of petroleum in the past three years has not dramatically increased the competitive potential of the unconventional sources of energy, at least in their traditional forms. And, as things are, the radically new technologies suggest no ways in which the sun or the tides or the interior of the earth might make substantial contributions to the world's energy economy before the end of the century.

This judgement, not everywhere accepted, turns on the

nature of technological development. In the past few years, the technical literature has been filled with schemes for making use of solar energy among other unexploited naturally occurring sources of energy. For the most part, however, the schemes suggested remain undeveloped. Moreover, each of these unconventional sources of energy suffer from important technical and thus economic disadvantages. For one thing, the working temperatures of many of the devices suggested are low, which means that the efficiency with which, for example, the energy might be converted into electricity is limited. But most of them also have the disadvantage that the source of energy is not concentrated physically, which means that many of them are able to yield from a single installation much less energy than would be produced from, say, a large modern electricity generating plant. Moreover, it is unlikely that progress in the development of these new sources can be rapid, from which two conclusions follow. First, no amount of research and development in the next few years can make a substantial difference to the degree to which they augment the supply of energy this century. Conversely, in planning for the period immediately ahead, research and development on other efforts is likely to yield practical results more quickly.

The potential of solar energy is a vivid illustration of the kinds of difficulties to be overcome. There are at present several ways in which the sun's energy might be turned into more usable kinds of power. For several years, various experimental projects have used large mirrors to concentrate the sun's energy – the French government, for example, has supported projects in the Pyrenees and North Africa. Technically, it is feasible to use the energy thus obtained to produce steam, which may in turn be used for driving electrical turbo-generators, or to produce electricity directly by devices essentially similar to the solar cells which provide earth satellites with electricity. Others favour schemes in which solar energy is used to support the growth of bacteria and other organisms which in turn may yield materials, such as hydrogen, which could be burned as fuel.

Whatever process is suggested for making use of solar energy, however, the daunting truth is that, when account is taken of the amount of solar energy absorbed in the atmosphere, and of the likely efficiency with which solar energy can be converted into electricity (probably 10 per cent at the most), the area of the earth's surface needed to produce 1000 megawatts of electricity is likely to be at least 42 square kilometres, or roughly 16 square miles. What this implies is that 1 kilowatt of electrical generating capacity corresponds to 42 square metres of the earth's surface, and that, if this electricity is to be competitive with other forms of electricity (nuclear power, for example), the investment cost should be no more than $250 per kilowatt; which in turn implies that whatever device is used for turning solar energy into electricity, the cost of collecting the sun's radiation could not be more than $5.00 per square metre, or rather less than the cost of covering the floors of domestic dwellings with the cheapest carpets.

For such reasons, it is hard to imagine that solar energy can become an important source of power for modern industry in the distant let alone the foreseeable future, which is not to say that there may not be important applications of new techniques in special circumstances. Solar energy can provide power for equipment installed in remote parts of the world. In places where people live, the application of engineering techniques to the design of buildings may trap and make use of some part of the large amount of solar energy at present reflected from buildings, though the extent to which techniques like this will in future be applied will depend critically on more thorough assessments of what is economically possible.

So the most potent uses of solar energy in the decades ahead are likely to be found in agriculture, either in the improvement of the efficiency with which plants convert solar energy into chemicals (not just for food) and in the development of crops which might replace chemical manufacturing processes dependent on fossil fuels. Thus some of the agricultural research now under way suggests that it may be possible to develop plants which are much more efficient at converting atmospheric nitrogen into nitrogenous chemicals, which might make it possible to reduce considerably the present consumption of artificial fertilizers in agriculture, saving energy from fossil fuel to the tune of perhaps 100 million tons of oil a year. By the same test, it is possible that, with the increased prices of petroleum, alternative routes towards what are at present called petrochemicals might become not merely possible (which they are at present), but also economic. The use of sugar from sugar cane as a means of producing a wide range of chemicals has been recognized for several years, and could, if the prices were right, mean the elimination over a period of decades of a fossil fuel consumption (mostly as petroleum hydrocarbons) already equivalent to some hundreds of millions of tons a year.

Solar energy is also responsible for wind, rain and various of the temperature gradients in the earth, all of which have been suggested as possible sources. And, indeed, in many parts of the world, there are still great opportunities for increasing the production of hydroelectric power. It has been calculated that the total potential for generating power of the world's streams and rivers is nearly 3 million megawatts, the equivalent of 4500 million tons of oil a year, or nearly as much as the world's present consumption of energy. But hydroelectric plants do consume large quantities of land beneath the artificial lakes established upstream of the impounding dams, produce electricity intermittently with seasonal changes of the local weather, unless they are very large, and are usually a long way from the places where electricity is needed. One quarter of the potential reserves of hydroelectric power are in Africa, for example. Thus, although hydroelecticity is a small but steadily growing part of the energy production in advanced industrialized societies, only about 5 per cent of the potential resource is at present used.

Windmills, widely used in the Middle Ages, are also probably capable of further development now that new technologies are able to make their sails more efficient by techniques learned from the design of aircraft, and to construct much larger sails than would have been thinkable in medieval times. In the United States, as well as in European countries such as Denmark and Hungary, large modern windmills have been operated successfully. A windmill for generating 1250 kilowatts of electricity was installed in 1941 on a hill called Grandpa's Knob in Vermont. More recently, the National Aeronautics and Space Administration has embarked on a more advanced scheme of this kind, but there is as yet no sign that the capital costs of generating electricity from devices like these will make them economically worthwhile. And there is, in any case, a limit to the use of windmills for generating electricity on a large scale: wind is intermittent, energy must somehow be stored, and large numbers of windmills placed to make use of more than a small fraction of the wind energy reaching a particular region would have profound but incalculable meteorological consequences.

In the present search for new ways of making energy, there is only a thin line between science fiction and reality. Some have suggested that large quantities of solar power might be captured by placing vast arrays of solar cells in orbit around the earth and then transmitting the power back to where it is needed by microwaves – a scheme that begs a great many questions about several still immature technologies. It has also been suggested that energy in very large amounts might be extracted from the process which normally occurs in river estuaries, where the mixing of fresh water with salt water is theoretically the equivalent to the energy released if the river tumbles into the sea over a dam nearly 300 metres high. Unfortunately, to extract such a large amount of energy it would be necessary to pass the whole of the flow of a river through a membrane, which is technically about as easy as passing a camel through the eye of a needle.

Of all the futuristic schemes for extracting energy ultimately derived from the sun from sources not so far even considered, let alone exploited, the most hopeful is that which depends on the difference of temperature between the surface of the tropical oceans and water lying a thousand feet or so down. On paper, again, it should be possible to use this temperature difference, constantly maintained by the heat from the sun in the tropics, as a source of electricity. All that is necessary is to construct a machine that would function as a refrigerator in reverse, designed to float at the appropriate level in the sea. Here, again, one difficulty is that the scale and intricacy of the machinery means we can as yet do no more than guess at the cost, let alone estimate the climatic consequences or the cost of transporting the energy, either as electricity or as hydrogen gas, to where it was needed.

Whatever its disadvantages, solar energy at least has the virtue of not being a source of pollution, which is no doubt one of the reasons why it has become the darling of the environmental movement. In reality, any worthwhile application of solar energy in the years ahead is likely to create such serious problems in the use of land or sea as seriously to disturb the localities in which it was produced. Even growing vast acreages of sugar cane (which would require large amounts of irrigation water in most suitable locations) would to many people be offensive.

And much the same, of course, is true of schemes for extracting power from the tides, ultimately derived from the gravitational attraction between the earth, the sun and the moon. In France, tidal power has been produced since 1966 to the tune of 240,000 kilowatts by means of a dam punctuated by turbines on the estuary of the River Rance in Britanny. A similar installation has been operating in the Soviet Union. But schemes for exploiting the tidal energy in the Bay of Fundy, as well as in the River Severn in Britain, have until now been considered uneconomic. The quantities of energy to be obtained are small even compared with the production of hydroelectricity from most hydroelectric dams of similar size, and there is every likelihood that tidal power can be only a marginal source of power, and an expensive one at that, except where both tide patterns and geography provide favourable circumstances.

Geothermal energy has a brighter future but, despite frequent claims to the contrary, it is potentially an environmental nuisance. So far, the exploitation of geothermal energy has been confined to those parts of the earth's surface where underground steam can be extracted with comparatively little engineering preparation and used to drive electric turbines. In northern California, for example, at a site known as the Geysers, the Pacific Gas and Electric Company produces 180 megawatts of electricity at a total cost rather less than that from conventional electricity generating plants. Similar devices are installed in Italy, New Zealand, Japan and the Soviet Union, and by 1972 the total generating capacity of these installations was nearly 1200 megawatts. During 1974, the Department of the Interior in the United States put up for lease several tracts of territory on either side of the Sierra Nevada where there is good reason to think that subterranean steam may be found within about a thousand feet of the surface.

Unfortunately, it is not yet known how long the supply of steam from an underground reservoir will last, though the Pacific Gas and Electric Company hopes eventually to obtain between 1000 and 4000 megawatts of electricity from the Geysers field. Moreover, subterranean steam also contains large quantities of hydrogen sulphide, a foul-smelling gas which must either be released to the atmosphere or removed. Geothermal sources of steam may also yield large quantities of water saturated with chemical salts which cannot be disposed of simply. Given the comparatively low temperature with which geothermal steam is usually produced, the efficiency of the electricity generating plant is necessarily low, usually less than 25 per cent, and so comparatively large amounts of waste heat are discharged into the atmosphere.

With luck, some of these difficulties may be avoided by exploiting other sources of energy in the interior of the earth, usually at greater depths. Thus it is recognized that in many parts of the earth's crust, particularly in places such as the mountainous regions in the western United States, potential volcanic activity at depths of between 5000 and 10 000 feet accounts for the presence of large masses of hot rock. There are now several schemes for the development of such sources of energy, often by the use of a deep underground explosion of a nuclear device to create fissures from which water pumped in from the surface might extract the heat.

Certainly there is no reason why underground reservoirs of heat like these should not be continuously exploited over half a century or so, but the technical difficulties remain to be assessed. Because of the tendency for underground water and steam to contain large quantities of chemicals, there is a danger that the turbines of electricity generating sets may be damaged by corrosion, though turbines using hydrocarbon gases in a closed cycle might be used instead if they proved practicable. It is also significant that even the most optimistic estimates of the potential contribution of geothermal energy in the United States suggest that, by the end of the century, the equivalent of 150 million tons of oil a year might be produced in such ways. What this implies is that geothermal energy, potentially cheap but with its future pace of development technologically uncertain, is worth having, but is unlikely to do more to bring supply and demand into balance, even in the most geologically well-endowed part of the world, than would, for example, the reform of United States policies on the pricing of natural gas.

The bearing of these technical considerations on the supply of energy in the decades ahead is clear. With some exceptions, the technology of extracting energy from these unconventional sources is not sufficiently developed for their economic potential to be accurately assessed. In most cases, as with geothermal and tidal power, the chances are that the quantities of potentially usable energy are small compared with present world consumption. And in some cases it is clear that costs are likely for the foreseeable future to be very high.

So why is there such enthusiasm for these projects? The belief that geothermal energy is somehow 'clean' is technically mistaken, while the belief that solar energy, however it is exploited, could avoid environmental nuisances can only be held by those who exclude the use of land from their criteria of a public nuisance. It is true that these sources of energy are 'renewable' in that in most cases they involve no irreversible consumption of resources – one notable exception is hydro-electric power, for even the best dams silt up. To be set against this, however, is not merely the cost but the fact that, except in special circumstances, these alternative sources of energy are at present not potential contributors to the world's supply of energy. By the turn of the century, they may be important. In the 1970s, they are still things of the future.

9
Power for coming decades

In the 1930s, when world demand for energy was growing rapidly, if not as quickly as later, when nobody could be sure that the Middle East would turn out to be a prolific source of cheap petroleum, and when the coal industries, the chief source of energy in industrialized countries, were showing signs of strain, it would have been entirely appropriate for governments to have taken fright about the threat of an impending scarcity of energy. For some reason, possibly because of the world-wide recession of the 1930s or the threat of the Second World War, they were not. That is a puzzle for historians.

By the time that justifiable concern about the future of energy supplies had surfaced, in the early 1950s, it had become plain that nuclear power, in some shape or form, would in due course make a substantial contribution to the world's energy economy. And from the mid-1950s, it became clear that nuclear power would ultimately be competitive with previously conventional sources of energy. In just over a decade, the outlook for the industrialized world was transformed more radically than even by the steam engine. It is not easy to understand why the lesson has not sunk home.

That the prospect of nuclear power, for military as well as civilian purposes, should have come as a bolt from the blue in the years immediately before the Second World War is a measure of how diffident were the ways of scientists in the 1930s and earlier. In 1905, Einstein provided an explicit arithmetical formula to show not merely that matter can be converted into energy, but also to calculate the energy that might be obtained in such a process. Moreover, from that time it was clear that mass should be a prolific source of energy. If it were possible to convert one kilogram (roughly two pounds) of any substance into energy, the result would be the equivalent of 2.15 million tons of oil. As yet, there is no practical means of converting mass into energy on such a scale. Nuclear energy is, for the time being, only a small step in that direction.

By the 1920s, it was known that atoms of different kinds, oxygen atoms or iron atoms or uranium atoms, are so constructed from atomic particles that the mass of each is less than the combined masses of the particles of which they are made. The whole is less than the sum of the parts. The difference is a measure of the energy that would be released if atoms (or, more strictly, atomic nuclei) were put together from their elementary constituents, as has happened in rapidly evolving and exploding stars. Nuclear energy is that produced by converting atomic nuclei, whose construction has released comparatively little energy, into nuclei which are, in terms of the energy the nuclei contain, more stable.

Hydrogen atoms are the simplest of all, uranium atoms the most complicated. Atoms which lie in between, such as iron atoms or strontium atoms, have masses which are somewhat less than would be expected from a simple interpolation between the two extremes. What this implies is that if it were possible to divide the nucleus of a uranium atom into halves, the result would be two atomic nuclei less massive than the starting material. As luck will have it, uranium nuclei will so divide, which is the basis of nuclear fission, the principle of the nuclear bomb and the nuclear power station. Conversely, it is also possible to make lighter atoms combine together to form atoms of intermediate complexity, and again the mass that disappears does not vanish into thin air but appears as energy. Nuclear fusion is what provides the stars with energy and makes hydrogen bombs possible.

In the present climate of technology, it is unthinkable that such an important set of principles would be widely known, and even taught to students, without stimulating a search for some way in which they could be exploited. In the 1930s, however, the pace of innovation was slower and innovators were less ambitious. Although it was demonstrated as early as

1932 that atomic nuclei can indeed be broken up by the impact of atomic particles, to begin with protons, it was not until the closing years of the decade that the physicists recognized that atoms of uranium can relatively easily be broken into pieces roughly half as massive. The most convenient way of doing this is to arrange for a collision between the nucleus of a uranium atom and an atomic particle called a neutron. And as luck will have it, when a uranium nucleus is disintegrated in this way, there may be anything between two or three neutrons left over, whence the chain reaction that makes uranium atoms of the kind most susceptible to fission (uranium-235) into a nuclear explosive.

Whether the outcome of the wartime Manhattan Project is counted a success or a disaster, it was, long before nuclear weapons were dropped on Hiroshima and Nagasaki, plain that the systematic conversion of uranium into lighter elements could be carried out to release energy in usable forms. This, after all, was how the Manhattan Project produced plutonium for the Nagasaki bomb. There are three different kinds of uranium in the naturally occurring metal. Uranium-235, the nuclear explosive, accounts for 0.7 per cent of it. Natural uranium also contains traces of uranium-233, which is as suitable for making nuclear explosives as uranium-235, but in practice is usually manufactured by nuclear manipulation of a quite different material, thorium-232. The bulk of natural uranium, 99.3 per cent of it, consists of uranium-238, which is not suitable for nuclear explosives because neutrons ordinarily do not stimulate fission but instead stick to the nucleus, which is then eventually converted into a material called plutonium-239, also usable as a nuclear explosive and a nuclear fuel.

The arithmetic of nuclear fission is straightforward. In principle, the energy that can be obtained by the fission of one kilogram of uranium-235 is the equivalent of 2000 tons of oil. Weight for weight, uranium is two million times as energetic as oil. The uranium-235 in a ton of natural uranium is energetically the equivalent of 14 000 tons of oil. But because uranium-238 can, in principle at least, be converted into plutonium-239, as potent a nuclear fuel as uranium-235, the long-term prospect is that each ton of natural uranium can in due course be made to yield the energy equivalent of two million tons of oil.

Although, ton for ton, uranium is even less plentiful on the earth's crust than conventional petroleum – the mineable uranium reserves of the United States, for example, are reckoned to be about 15 million tons, of which only some 10 per cent could be mined economically at present prices – the energy content of this uranium is twenty times greater than the petroleum reserves with which the United States was originally endowed, estimated at 100 000 million tons. This, crudely, is the sense in which fission has transformed the outlook for the supply of energy, lifting the horizons of the industrialized world by several centuries.

As with other potential sources of energy, however, the fact that the world's uranium reserves are likely in themselves to be a sufficient source of energy for the industrialized and developing world for many centuries to come is irrelevant. What matters is not how much nuclear fuel can, in principle, be mined and converted into energy, but the cost at which it can be done.

The experience of the past fifteen years has provided a framework within which this question can be answered. First, fission reactors are likely, at least until 1990, to be used exclusively for the manufacture of electricity. Secondly, there are two kinds of fission reactors, known as thermal and fast reactors. The fuel in thermal reactors can contain quite large proportions of uranium-238 without bringing the nuclear chain reaction to a halt, but the penalty is that only comparatively small amounts of plutonium are produced as a by-product (typically, between a third and two-thirds as much plutonium as the uranium-235 consumed). Fast reactors, on the other hand, require a nuclear fuel that consists almost entirely of fissile material, either uranium-235 or plutonium-239, but they can be used not merely as sources of energy but also as means of converting uranium-238 to plutonium, in which case the amount of plutonium produced may exceed the amount of fissile material consumed, which is why these devices are sometimes called 'breeder' reactors.

The outstanding characteristic of nuclear reactors is that their capital cost is greater than that of conventional electricity-generating plants. Throughout the 1960s, when it seemed likely that uranium and conventional fuels would be marginally competitive, a great deal of ingenuity was lavished on attempts to make strict comparisons between the costs of nuclear and conventional plants, a largely fruitless exercise when costs were increasing rapidly and there was little experience of nuclear reactors in operation. In present circumstances, it is reasonable to assume that the capital cost of a nuclear power station built to one of the three patterns now available commercially in the United States is between $600 and $700 for each kilowatt of generating capacity, or roughly a third as much again as the capital cost of oil-burning generating plant.

Nuclear power stations are more capital intensive when compared with other investments in energy resources. Their advantage is that the cost of the fuel is low, and that the overall cost of the electricity they produce is relatively insensitive to the cost of uranium. At the end of 1973, for example, the Philadelphia Electric System in the United States convinced itself that nuclear power would be cheaper than electricity from oil-fired power stations so long as the cost of fuel oil was greater than $5.00 a barrel. But fuel oil was already costing more than $6.00 a barrel, so that the utility's decision to build nuclear power stations (in this case, the High Temperature Gas Cooled Reactor, or HTGR, designed by Gulf-General Atomic) was commercially prudent. Indeed, it is not surprising that the rate at which new nuclear plant has been ordered in the United States accelerated rapidly during 1974.

Through most of the 1960s, only small amounts of nuclear generating capacity were ordered in the United States, although in the three years 1966–8, when Westinghouse and General Electric (the manufacturers of the pressurized water reactor and the boiling water reactor respectively) were offering nuclear plants on a fixed-price basis which experience has shown to have been unprofitable, American utilities ordered a total of 60 million kilowatts of generating capacity, the equivalent of close on 120 million tons of oil a year. At the end of 1973, nuclear generating capacity in the United States amounted to 25 000 megawatts (1 megawatt = 1000 kilowatts), but a further 53 000 megawatts of capacity was either ready to come into commission or under construction. In

1974, new generating capacity was being ordered at the rate of 30 000 megawatts or more a year, and the rate of addition to the United States nuclear generating capacity may even be greater in some succeeding years.

Estimates of the total amount of nuclear capacity installed in the United States by 1985 vary considerably. In 1973, the Atomic Energy Commission estimated that there would be 280 000 megawatts of nuclear generating capacity in 1985, but more recent studies suggest that total capacity by 1985 may amount to 305 000 megawatts, the equivalent of between 500 and 600 million tons of oil a year.

Outside the United States, in Western Europe and Japan, the pace of change has been slower, though the government of France, early in 1974, took the decision that all future additions to the electrical generating capacity of France would consist of nuclear reactors and that electricity itself would account for a growing proportion of the country's energy consumption. In Britain, which during the 1960s ordered new nuclear generating capacity more rapidly than any other industrialized community, the total generating capacity at the end of 1974 was the equivalent of 8 million tons of oil a year and will be increased to 12 million tons of oil a year when nuclear power plants then under construction come into service. But there is likely to be a hiatus in the commissioning of new plant in the remainder of this decade, and it is too soon to tell how quickly the development of the steam-generating heavy water reactor chosen by the British government in 1974 as the design for future thermal reactors will make a sizeable contribution to the country's energy supplies. During 1974, the International Atomic Energy Agency demonstrated, on the basis of a careful market survey, that even in developing countries, whose electricity generating systems are able to assimilate only comparatively small nuclear reactors (between 100 and 400 megawatts of electrical capacity compared with reactors of capacity greater than 1000 megawatts now considered economic in the United States), thermal nuclear reactors would be economically advantageous if oil prices are greater than $3.50 to $5.50 a barrel, a lesson that many developing countries will be unable to take to heart for lack of capital.

The combined effect of the plans now laid on the world's supply of energy by 1985 will be substantial. Even allowing for the long construction time of nuclear power plants (between six and nine years), and the possibility that some of the plans now laid may not be executed in precisely the way intended, nuclear power stations are certain to be yielding energy the equivalent of 600 million tons of oil a year by 1985 and possibly as much as 1000 million tons. In other words, the total amount of electricity generation from nuclear reactors will be the equivalent of something between the total production of petroleum in the United States and that from the Persian Gulf at present. This is the sense in which it can be held that the promise of nuclear power is already a reality. The strategic importance of this rapid growth of the nuclear industry in the dialogue between oil consumers and oil producers is plain. Indeed, if they are clever, the major energy consumers will be able to establish the point that the price of energy of all kinds should be determined by the cost of generating nuclear electricity. That, at least, should be their objective.

This is why the technical problems which at present beset the nuclear industry need more urgent attention than they have been given. The scale of the growth now planned dwarfs any previous transformation of industry. The plans now laid in the United States, for example, entail the design and construction of new reactors at the rate of two or three a month for the next decade. The steel fabrication industry will somehow have to find a way of manufacturing thirty reactor pressure vessels a year. No fewer than 30 000 to 40 000 people will have to be recruited and trained as operators of nuclear plant and maintenance personnel. The development of the steam engine industry in the nineteenth century took half a century. The exploitation of a technology every bit as novel is now planned in a mere decade.

Some of the obstacles to change are administrative. In the United States, it is estimated that the licensing process to which all nuclear plants on new sites are subjected takes an average of three years – two years for the regulatory agencies to analyse the plans submitted, and then a period of between a few months and two years during which public hearings may be held. Although the detailed design of a new plant may continue during this period (with some financial risk), so that construction can begin once a construction permit has been issued, there are the strongest reasons for an energetic effort to shorten this procedure. If by attention to licensing procedures and the mechanics of constructing nuclear plants it were possible, in the United States, to shorten the construction time for reactors by between two and three years, the result would be an extra amount of energy comparable with the production of petroleum from the North Slope of Alaska.

The certainty with which the nuclear industry can grow in the decade ahead will, of course, be determined by the practical skills of the manufacturers, and their record in the past fifteen years has not been encouraging. A part of the trouble has been the irrational expectations of governments as well as the technical community in the past thirty years. The truth is that the development of the technology of civil nuclear power has been a task of great complexity, and should have been recognized as such from the outset. The speed with which nuclear weapons were manufactured during the Second World War was an entirely misleading guide to the difficulties involved in the development of nuclear power stations, and the lighthearted optimism of the early would-be designers of nuclear power stations is not easily excused even with the passage of twenty years.

In retrospect, it is also a great misfortune that the energies of those involved in the technical development of nuclear reactors have been spread over a host of radically different designs. Where diversity might have meant an accurate identification of the most economical ways of producing nuclear power, it has instead meant that the efforts of the world's technical community have been too thinly spread. The failure of European nations to agree on an effective programme of nuclear power has done serious damage. The research and development programme launched within Euratom in 1958 consisted largely of projects the then member nations did not want to pursue on a national basis, for the sake of the illusory commercial advantages of national success. Thus the Euratom programme collapsed within a decade.

In the circumstances, it is easy to understand why the brief history of commercial nuclear power is littered with spectacular engineering failures. The Enrico Fermi reactor near Detroit (intended as a prototype of a commercial fast reactor) was first put into operation in 1966 when the technology of fast reactors was still poorly understood. Within a few months, the reactor was out of commission because two pieces of zirconium became loosened. Though it is often said that the 1.5 million people of Detroit were put in hazard, the safety procedures had functioned as intended. However, the reactor remained out of commission for four years, by which time it had become clear that its design could contribute little to the development of commercially viable fast reactors.

As bad luck had it, the successor to the Enrico Fermi reactor, a fast reactor planned for a site on the Clinch River in Tennessee, was still in trouble in 1974. Although essentially similar in design, the final cost of this reactor is likely to be at least twice the first estimate of $700 million, with construction at least two years behind.

In Britain, the accident in October 1957 at one of the two reactors originally built at Windscale to produce military plutonium cannot strictly be called a failure of the civil nuclear power programme, for the design of the reactor had no similarities to civil designs. The release of radioactivity that followed the accident was a foretaste of the kinds of accident that could occur at nuclear plants, which is variously interpreted as threatening or reassuring. More serious if less of a public hazard has been the discovery that the reactors built in the first phase of the British nuclear power programme (using natural uranium as fuel and carbon dioxide to extract heat) cannot be operated at full power, partly because the graphite used for slowing down neutrons interacts with the cooling gas, and partly because mild steel bolts used in the cooling system are corroded by it.

A second generation of gas-cooled reactors, ironically known as advanced gas-cooled reactors, has been less successful still, partly because the design of full-scale reactors was undertaken with only limited experience with a prototype reactor a tenth the size, and partly because a misguided attempt to put British industry on its competitive mettle encouraged a needless diversity of detailed design. And elsewhere, as in France, Italy and Japan, reactors have operated below their design power, or less continuously than was originally intended, because the failure or threatened failure of quite trivial components was not anticipated at the outset.

What this record has shown empirically is what has been known for thirty years: that nuclear reactors differ from most other machines devised by modern technology in that, once built and operated, they cannot be seriously modified or repaired throughout their operating life. Radioactivity prevents it.

With this background of comical misadventure, why should it be sensible for industrialized nations now to trust in nuclear power for a strategically important contribution to energy supplies in the 1980s? There is, of course, no objective answer. Nobody can be sure that reactors now being designed will function in every way as their designers intend. But running through the developments of the past fifteen years is a thread of technical success and competence that cannot be discounted. The great majority of the reactors now operating *are* working as their designers intended. Departures from design performance can be understood. And the past two decades have brought not merely increased competence, but a sense of sobriety into the craft of nuclear design.

Whether the present pace of growth in the nuclear industry can be sustained is another matter. Most types of thermal reactors use, in their fuel, uranium enriched in the fissile isotope uranium-235 to between 2 and 5 per cent (compared with the 0.7 per cent of uranium-235 in natural uranium). Although plants for producing enriched uranium by the techniques adopted in the Manhattan Project are being built in the United States and France, while a consortium of British, Dutch and German public authorities is embarked on producing enriched uranium by using centrifuge machines, there is a risk that these facilities will not be sufficient to process all the nuclear fuel needed in the late 1970s and early 1980s. Here, then, is a potential bottleneck that can only be removed if steps are taken soon.

Much the same is true of the potential bottleneck at facilities for reprocessing uranium fuel, a process in which fuel discharged from nuclear reactors must be treated chemically so as to separate unused uranium, plutonium and the radioactive materials which result. In spite of a decision by the US Atomic Energy Commission in 1960 that the reprocessing of fuel should be made a strictly commercial operation, none of the three commercial plants on which work began was functioning in the second half of 1974. At that time, the only separation plants in the world (outside the Soviet Union and mainland China) which could reprocess commercial nuclear fuel were those in Britain and France. This deficiency will not be remedied until the end of the 1970s. Shortages of reprocessing facilities are serious because delays in reprocessing spent fuel have serious economic consequence.

Of all the constraints on the development of nuclear power in the coming decades, however, the restricted availability of uranium is likely to be the most powerful. As with other fuels, what matters is not how much there is in the ground or the sea, but how much is available for use, and at what price. Although the potential reserves of uranium are very large, the rapid growth of the nuclear power industry now in prospect is likely to create intense competition for accessible supplies while straining the capacity of the world's enrichment plants. The combination of these pressures is likely, in turn, to strengthen the case first for the development of fast reactors that are more efficient than thermal reactors at converting uranium to plutonium (and which are, in any case, the most convenient ways of using plutonium as fuel); secondly, for the use of thorium as an alternative nuclear fuel (which, in turn, will depend on the development and probably, in practice, on the accumulation of plutonium stocks as well); and thirdly, in the more distant future, for intensifying the search for practical methods of winning power from fusion rather than fission.

Only time will tell precisely how quickly the demand for uranium will increase in the years ahead. Much depends on the kinds of reactors that are built, though the chances are that the annual demand for natural uranium will have grown from 1900 tons in 1970 to 30 000 tons in 1975 and to between 1000 000 and 150 000 tons a year in the second half of the 1980s.

Three things are clear. First, while plenty of uranium exists in deposits that can be mined economically at current prices, there is a serious danger that it may not be mined quickly enough to satisfy the demand expected in the 1980s, when generating plants now being ordered come into service. This was one of the anxieties underlying the research and development programme for the United States put forward at the end of 1973. A part of the trouble is that the uranium mining industry, in the doldrums for much of the 1960s, is not able to respond quickly to the need for change. Another difficulty is that detailed exploration of known deposits is ill-prepared, while there is a possibility, especially in the United States, that there will be environmental objections to the extraction of small proportions of uranium oxide from the very large quantities of rock that must first be mined and afterwards disposed of.

Secondly, because the distribution of uranium in the world is no more closely related geographically to the places where it will be used (except that the United States is well supplied) than is the geographical distribution of petroleum, there is a danger that in the years ahead the price of uranium may be linked with the current artificially high price of other forms of energy and, in particular, by the OPEC price of oil. The moral is that potential consumers should act quickly, in their own and everybody else's interest, to increase the supply of uranium while making the business profitable for those who mine it. It is ominous that the Australian government, with one of the potentially largest sources of uranium ore, declared in October 1974 a two-year moratorium on the development of new uranium deposits. The situation is comparable with that exploited by OPEC, and some of the largest suppliers of uranium – South Africa and Australia in particular – will, for perfectly understandable reasons, do everything they can to increase their share of the economic rent. This, no doubt, is the chief reason why the South African Atomic Energy Authority has embarked on a plan to produce enriched uranium.

The third and most important consequence of a potential physical shortage of nuclear fuel is that there is an urgent need for the development of fast reactors to be carried to the point at which they could, if necessary, be built without delay. The issues are complicated, and raise commercial as well as technical considerations.

In all nuclear reactors, even those which use natural uranium as fuel, it is inevitable that some material which is not fissile, such as the uranium-238 component of natural uranium, should be converted into fissile material, such as plutonium-238. In many of the thermal reactors now in service, for each 100 atoms of uranium-235 consumed by nuclear fission, between 50 and 70 atoms of plutonium-239 are formed by the conversion of uranium-238. (Not all of these can be extracted when the fuel is reprocessed, for some of them will be used up in the fission process exactly as if they were atoms of uranium-235.)

In principle, there is no reason why thermal reactors should not be designed to yield more fissile material than they consume, or, in other words, to be 'breeder' reactors, and indeed the particular design of the high-temperature gas-cooled reactor offered by the Gulf-Shell consortium in the United States (the first version of which was commissioned in 1974 at Fort St Vrain in Texas) is intended eventually to use uranium-233 as fuel and to convert thorium into uranium-233 more quickly than the original charge of fuel is consumed. But the efficiency with which non-fissile materials such as uranium-238 can be converted into nuclear fuels is much greater in fast reactors – reactors in which no deliberate steps are taken to make the neutrons travel slowly. Prototype fast reactors, using liquid sodium as a means of extracting heat, were commissioned during 1973 and 1974 in France and Britain, with a very similar device following closely behind in the United States. It appears to be possible, on the basis of these small reactors, to design full-scale (1000 megawatt) reactors in which the amount of plutonium produced as a by-product of the fission process is between 10 and 20 per cent more than the amount of fissile material consumed.

For the time being, however, little is known of the cost of electricity generated in fast reactors, although in the long run (in the 1990s) the capital cost of building fast reactors should not be very different and may even be less than the cost of building thermal reactors. Given that fuel costs will of necessity be less, fast reactors are plainly likely to be commercially preferred. In the long run, however, the significance of fast reactors extends beyond the day-to-day cost of producing electricity. They are strategically essential if the cost of uranium is not to be sharply increased.

This promised land, it must therefore be acknowledged, is still some distance away. Although the potential of the fast reactor as a means of producing more fuel than it consumes was recognized in the 1940s, and in spite of tangible steps in this direction in the United States and Britain in the 1950s, it is unlikely that electricity utilities will be able confidently to order commercial reactors of this kind before the early 1980s, and therefore unlikely that fast reactors will make substantial contributions to the production of electricity even in the most advanced communities before the end of that decade.

There are several explanations, some technical, some of a more general kind. But the plain truth is that in most places where nuclear power is technically advanced, in the United States, Western Europe and possibly the Soviet Union, the effort expended on fast reactors during the 1960s was much less than was necessary to carry through unavoidably long-term programmes. The illusion of the 1960s, that cheap oil would remain plentiful, accounts for much of the thoughtless curtailment of the programmes for the development of fast reactors in countries such as the United States and those of Western Europe. But very little can be advanced as explanation of the way in which at least four countries – France, Germany, the United Kingdom and the United States – have independently pursued essentially similar programmes of development, each of them inadequate.

Technically, nuclear power is as much one of the wonders of the modern world as the enthusiasts of the 1950s believed. Nuclear power stations will, in the years ahead, provide much of the need for energy that would otherwise have had to be met by burning oil, and they will, from the early 1980s, increasingly influence the cost of energy on the international market, provided the industrialized nations pursue the cause energetically. By the turn of the century, nuclear power will almost inevitably determine the commercial framework within which other sources of energy are used, and that situation will remain for several decades. So it is proper to acknowledge that nuclear power, like other new technologies, is not an unmixed blessing.

That nuclear energy carries novel risks has been recognized from the earliest days of the Manhattan Project. The issues are several, but all of them entail questions about the ways in which radiation, or radioactivity, might damage people or the environment in which they live. The problems are serious, and cannot be discounted. While there is no reason to suppose that the novel problems of radiation need be a serious restriction of the development of nuclear power in the years ahead, they do predicate the need for continuing vigilance, both in the investigation of the consequences of radiation for people and in the management of nuclear plants. The fact that the scientific issues involved are complicated, and that public understanding has been hampered (especially during the period of heavy radioactive fallout from nuclear weapons tests in the 1950s) by the complacency of the responsible authorities, has become, with the passage of time, a further source of confusion.

The facts are these. Radioacticity as such is not novel, even though the discovery of radioactivity and the recognition of its character dates back only to the closing years of the nineteenth century and the classic investigations of Bequerel, the Curies and Rutherford. Radioactive atoms are atoms whose nuclei disintegrate spontaneously, but by processes distinct from that of fission in which a nucleus splits, either spontaneously or by the stimulation of impact by a neutron, into two roughly equal halves. Naturally occurring radioactive atoms are of two kinds: some radioactive nuclei give off an electron, some give off the nucleus of a helium atom, called an alpha particle. In either case, the chemical nature of the atom is changed. Naturally occurring carbon-14 atoms, which give off electrons or 'beta particles' spontaneously, are in the process converted into stable atoms of the isotope nitrogen-14. Nuclei of uranium-238, which give off alpha particles, are converted in the process to nuclei of protoactinium-234, a relatively unstable nucleus which is quickly converted by the further emission of an alpha particle into the nucleus of radium-230. In radioactive transformations, the emission of alpha or beta particles is accompanied by the emission of energetic X-rays, called gamma-rays. Those atomic nuclei which are susceptible to spontaneous fission are also radioactive, emitting alpha particles. Many of the materials formed in the fission of heavy nuclei are themselves radioactive, but the emission of electrons or beta particles is predominant in the fission products. Occasionally, some unstable atomic nuclei emit positive electrons, called positrons.

Why should radioactivity be potentially harmful? The pernicious anaemia of which Marie Curie died in 1934 is one memorable proof of what can happen. For a start, alpha and beta particles are potentially damaging to the tissues through which they pass. So, too, are gamma-rays. Sometimes living cells can be killed. Sometimes their constitution may be changed, with the result that they may become cancerous or, if they are germ cells, the chromosomes they contain may be damaged in such a way that genetic defects are passed on to the offspring of the individuals exposed. These are some of the more spectacular of the biological consequences of nuclear radiations, a somewhat misleading term because only the gamma-rays are radiation in the strict sense. But to single these out is not to suggest that other biological consequences of exposure to radioactivity may not exist – experiments with animals have suggested that an animal's life span can be shortened by exposure, though the reasons remain obscure.

The arrival of the nuclear power industry based on fission, and the prospect that the industry will grow rapidly in the decades ahead, has increased the chance that people will be exposed to damaging amounts of radioactivity. So much is beyond dispute. At the same time, it is important to realize that all living things, now and in the past, are and have been exposed to quite substantial amounts of naturally occurring nuclear radiation. Cosmic rays are one source of exposure. So are the radioactive materials in the ground and those naturally occurring radioactive isotopes incorporated in the body. To suggest that living things have always been exposed to radiation is not to imply that the production of large quantities of artificial radioactivity in fission reactors has no significance, but it is proper that natural radioactivity should provide a yardstick to assess the biological consequences of artificial radiation.

So how safe is nuclear power? In the past few years, the prospect of a rapid growth of the nuclear power industry has stimulated many forbidding scenarios. In the 1950s, when much less was known of the biological consequences of radiation and the practical operation of nuclear power stations and their ancillaries, certain scenarios for catastrophe may have been forgivable. Now it is possible much more accurately to identify the points needing close attention. The overriding principle is that the safety of the nuclear power enterprise will depend critically on its success in preventing contamination of the environment with its by-products. And in spite of the regularity with which false hares are started, the points on which sober concern should centre can be identified with confidence.

First, nuclear plants, power stations and particularly the plants that reprocess fuel, routinely discharge to the environment small fractions of the radioactivity they produce. Liquid effluents, while they carry quite large amounts of radioactivity, are not a serious technical problem, for it is comparatively easy to identify the most sensitive means by which they might do damage, to lay down regulations to control the dose that people might then receive and finally to apply these regulations. The gases which nuclear power stations give off are, by comparison, a more serious source of hazard, partly because they are less easily monitored and controlled. Even in the brief history of the past twenty years, there have been several incidents in which a leaking container for a piece of uranium fuel has allowed radioactive iodine to escape. In reprocessing plants, the release of radioactive krypton (krypton-85) and radioactive hydrogen (called tritium) are, with present practices, unavoidable, and there is some cause to fear that if the nuclear power programme grows quickly between now and the end of the century, it may be necessary to change the present practice. Technically, this is feasible; administratively, it would be difficult – principally because the ubiquitous spread of radioactive materials in the atmosphere predicates the need for international agreement on codes of practice. And there is no easy way of legislating for failure. But routine contamination of the environment is at once a soluble problem and one to which nuclear power authorities and their regulatory agencies seem alert.

Secondly, there is the problem of managing the large quantities of radioactivity separated from spent fuel elements and consisting chiefly of radioactive fission products mixed with plutonium in small proportions as well as other artificial elements similar both chemically and in their nuclear properties. On the face of things, the management of radioactive wastes is a bizarre and unprecedented technical problem, principally because some of the radioactive materials in the fission products are exceedingly persistent.

In the nature of things, the radioactive process is self-destructive. Because radioactive atoms are transformed in the act of emitting nuclear radiation, radioactivity vanishes in the course of time. The snag is that the times vary enormously. For example, half of any quantity of the radioactive material iodine-131 produced in nuclear fission disappears every eight days, a further half in the next eight days, and so on. Some important by-products of the fission process have a still more fleeting existence, but, at the other end of the scale, materials such as plutonium-239 are such that half of any quantity takes 24 000 years to be converted by radioactive decay into something else (in this case, uranium-235).

Considerations such as these dictate the strategy that must be followed in the management of radioactive waste. The problems begin with the discharge of spent fuel elements from reactors, and current practice at nuclear power stations require spent fuel elements to be stored six months in a pond of water for the radioactivity of the fission products to be reduced by natural decay. Thereafter, the highly active wastes are, for a time at least, stored in tanks, usually in concentrated form. There are technical difficulties stemming from the need to cool the tanks and to prevent sludge from collecting at the bottom. So long as the integrity of the tanks is not breached, storage is acknowledged to be indefinitely safe. To reduce the need for continual supervision, but also for economic reasons, most substantial processors of spent fuel rods have plans, likely to be implemented within a decade, for converting highly radioactive wastes into solid form, usually by the formation of boro-silicate glass. Further ahead lies the possibility of separating from the highly active wastes residual amounts of plutonium together with other artificial elements formed from plutonium and uranium by the action of neutrons, which together constitute the most persistent types of radioactivity from the operation of nuclear reactors. If it should be possible to implement this process, the most daunting aspect of the waste-storage problem should be removed, for it will then be possible to dispose of the long-lived radioactivity by returning the materials to a fast reactor, where they would function as a nuclear fuel.

One of the characteristic features of the waste-management problem is that the procedures now being followed are acknowledged as only temporary solutions. By the end of the century, it would not be feasible to store these highly active wastes in a series of tanks, however well built. So may it not be held that the nuclear industry is behaving fecklessly in relying now on a technique that must eventually be replaced? The validity of the charge turns on the confidence with which it is possible to predict that more permanent means of storage are feasible. And the research of the past decade has demonstrated clearly that long-term safe storage of highly active radioactive wastes is practicable. Indeed, there is a case for thinking that too rapid a development of the glass-making process may have the unfortunate consequence of postponing the point at which benefits might be won from the separation of the long-lived radioactivity.

The third set of safety problems concerns the occupational health of workers at nuclear plants. Two separate issues arise. Will the doses of radiation received by nuclear workers significantly damage their health, and will their accumulated dose of radiation create such a load of temporarily concealed genetic mutation that the health of future generations will be jeopardized? In the first case, the radiological protection standards now applied in nuclear plants are probably stringent enough to ensure that overt death rates are indistinguishable from the normal. (The fact that actual death rates in British nuclear plants are lower than in the rest of the population is more a tribute to the health restrictions on the selection of nuclear workers than a proof that radiation is good for people.) The genetic risk is more serious, if only because of the difficulty of knowing what the long-term consequences could be. And where the health of future generations is concerned, it makes no difference, strange though it may seem, whether a large dose of radiation is received by a small group of workers or a smaller dose is spread out among the whole population. In Britain in 1970, the genetic dose from occupational exposure was twenty times as great as the genetic dose to the population at large from environmental contamination with the radioactive by-products of the nuclear power industry, but, for what it is worth, rather less than one-twentieth of the genetically significant dose from the medical use of X-rays and 0.7 per cent of the genetic dose from cosmic radiation. At present, in other words, it is exceedingly unlikely that radiation is storing up a measurable, let alone an intolerable, load of genetic defects in the general population as a result of occupational exposure.

The fourth and most serious cause for concern about the safety of nuclear plants is the possibility of accident. The danger that a nuclear plant might explode like a nuclear weapon is unreal, but there are many other mishaps that could occur. Punctured fuel elements can, in some circumstances, lead to fires, such as that which occurred at the plutonium-producing plant at Windscale in Britain in 1957. In other kinds of reactors, interaction between the material of which fuel is made and a liquid, like water or sodium, used to carry away heat, can release large amounts of energy and create conditions within reactors when the rate of output of nuclear energy is increased. In these circumstances, the result of an accident will be the release into the environment of quantities of radioactivity whose nature will depend on both the character of the accident and of the reactor, and whose amount will be, by definition, a measure of the seriousness of the accident.

Both in the United States and Britain, the past decade has seen the development of systematic techniques for estimating the chances that reactor accidents of particular kinds will occur, on the basis of which it has been possible to develop criteria that must be satisfied by reactor designs. Briefly, in the United Kingdom, it is assumed that reactor accidents in which the quantity of radioactivity released is enough to kill between ten and 100 people should occur with a probability of one in 10 million or less for each year of a reactor's lifetime. Potentially more damaging accidents should be less frequent.

The design of reactors to such standards is feasible. The practical question is whether even these small risks are politically acceptable. For a reactor accident which caused between ten and 100 deaths would be one in which radioactivity affected between 100 000 and a million people, among whom deaths would occur at intervals of between ten and twenty years after the event. In the United States, largely because of the arguments of the Society of Concerned Scientists at Cambridge, Massachusetts, there has in the past few years been a vigorous public debate about the means by which American light-water reactors are provided with devices for flooding the reactor with water if something untoward happens inside. Essentially, the argument has been about the validity of the calculations by means of which the reactor designers and the regulatory authority have convinced themselves that they can indeed meet the criteria of safe operation they have set themselves. In this sense, however, the argument is in many ways beside the point. The overriding question is whether, with some hundreds of reactors in operation, and with the chance of an accident killing between ten and 100 people in any year thus being reduced to less than one in a thousand, governments and their electors would find the circumstance acceptable.

The conclusion that follows from this kind of calculation is simple. First, the continuing improvement of the reliability of reactors and their components is an important goal that can be achieved only when there is much greater experience of the way in which reactors function. Secondly, nuclear power on a substantial scale does predicate the need that those who hope to enjoy its benefits should also acknowledge that the risk of serious accidents, however small, is by no means negligible. The situation, in other words, is strictly comparable with that which attends many other recent technical innovations, like modern air transport. Although travel agents do not go out of their way to emphasize that he who purchases a ticket to travel by air thereby increases his chances of dying accidentally, the popularity of travel insurance is a proof that the travellers know the risks they run. Older industries, like coal mining, tend to concentrate their more obvious risks among those who work in them, but the calculation also takes into account the extent to which the general population is affected by air pollution. In short, and whatever may be accomplished by the reactor designers in the years ahead, the coming of nuclear power has brought with it a new kind of risk, comparable in many ways with that from other kinds of industries, but novel in that the damage might be done by radioactivity.

The reasons why populations of future decades will be expected to shoulder such new risks are easy to understand. If it were ever possible to make an abstract calculation of the benefits and risks of nuclear power, the benefits would far outweigh the risks. This is not, however, a situation in which abstract considerations are applicable. The plain truth is that in the advanced communities of the industrialized world, as well as in developing countries, the need to increase energy consumption is irresistible. It is unthinkable that communities such as the British would consider the unemployment that a sharp decrease in energy consumption would cause as more acceptable than the small risks of reactor accidents, just as it is politically unrealistic to expect developing countries to settle for slower economic growth or reduced agricultural output for the sake of keeping nuclear power at bay. Moreover, the rapid development of nuclear power offers the most promising route to the realignment of the cost of energy on which the economic welfare of the next ten years depends. The future, in a sense, is unavoidable.

10
The way ahead

A crisis is a turning point and, if perceived, a time for decision. That is what the dictionaries say. The events of the past few years, the physical shortages of energy in some industrialized countries and the increase price of crude oil decreed by OPEC during 1973 have required solutions to novel problems for many governments, and may thus in the short run be held to constitute a crisis. But they will not always seem like that. In the long perspective of industrial history, the last quarter of the twentieth century now before us will appear to follow naturally from those that have preceded it. Technologically, it will be the time when nuclear fuels supersede the fossil fuels. Historically, it will be a time when the benefits of industry are more widely shared between rich nations and some of those which have been poor. And if some states now rich should fail, through muddle, indecision and cowardice, to adapt to changing circumstances, that will be no surprise. Throughout the industrial revolution of the past two centuries, different communities have been justly rewarded for such good sense as they were able to command.

Long before the 1970s, it had become plain that there would have to be a radical transformation of the pattern of energy consumption in the industrialized world. Soon after the Second World War, it was clear that the coal industries of the industrialized West, the foundation of the first century of the industrial revolution, were a declining asset. Nothing that has happened in the past few years has changed that prospect radically. Coal is labour-intensive, and its cost will increase faster than the prosperity its use creates. Cheap petroleum, the foundation of the unprecedented growth of the 1960s, was always geologically a quickly wasting asset and has been made to vanish at a stroke of the pen by the decisions of the OPEC states in 1973. For the time being and for several decades to come, nuclear power based on uranium or thorium is the only practicable alternative, which is not to say that the use of the fossil fuels will now disappear. Rather, there will be a long period during which the new and the old can usefully (which means economically) coexist. Whether what are called future generations, our children or grandchildren perhaps, will find themselves exercised by the need to choose between thermonuclear power and solar power is an open question, and as yet an academic problem. What matters is, first, that there will by then be choices that can be made and, second, that the choices should be made rationally. With luck, the assaults on economic reality which characterized the 1960s and the hesitations of the early 1970s may be salutary admonitions.

Now, as for the past fifteen years, it is plain that the industrialized nations of the world have chiefly themselves to blame for the upheavals of the past few years. Severally and collectively, they have made errors which are economically and politically unforgivable. Above all, they have lost that blend of economic realism, social flexibility and capacity to welcome technological change on which the earlier decades of the industrial revolution were founded. Mercifully, one of the uncovenanted and still unrecognized benefits of the increased price of petroleum is that some at least of the industrialized nations will be restored to their former condition.

What are the mistakes? On what principles should they be rectified? There are several things to say. First, the continuing industrial revolution will be sustained, as it has been since the beginning, by energy. But there is no simple arithmetical rule relating even the growth of national prosperity, measured by economic statistics such as GNP, to the amount of energy consumed. Indeed, the experience of recent decades has shown that nations differ widely in this respect. In short, there is no reason why temporary or even permanent difficulties about the physical supply of energy, or increased prices for particular fuels, should necessarily impede the remarkable growth of the world's economy which has characterized the past two centuries. And, given that GNP is only the crudest guide to the well-being of a nation, there is no reason why the decades ahead should not be as productive of beneficent change as have been the decades past. The question that remains is whether the industrialized nations of the world, on behalf of themselves and the developing nations, will have the wit to chart a prudent course ahead.

Second, it is now plain, as it has been for the past fifteen years, that the management of energy supplies by the industrialized nations of the world has too often been, in economic terms, an attempt to make water run uphill. The United States has been the most serious and persistent offender, with its attempt to protect its domestic petroleum industry from overseas competition and with its vast apparatus of regulation of the electricity and natural gas industries designed to protect consumers from the economic costs of comfort and freedom to travel. The other industrialized nations of the West are guilty of a different kind of error. Throughout the 1960s they chose to avert their eyes from two inescapable truths about the supplies of petroleum from OPEC countries and particularly from those of the Persian Gulf – the geological fact that the reserves of even these enormously productive oil fields must be small compared with other potential sources of energy, and the political certainty that, sooner or later, the oil-producing states would be in a position to claim a fair share (which means the whole) of the economic rent from the exploitation of their petroleum reservoirs. With the impetuous entry of the United States into the international petroleum market in 1970, that time had arrived. What has happened since has followed inexorably.

One of the paradoxes from which the disruption of the energy market has sprung is that the industrialized nations of the West, which include some of the most vigorous exponents of the doctrines of the free market economy, have consistently behaved as if sources of energy should be exempt from the rules of Adam Smith. Two contradictory distortions of economic principles have been engineered, sometimes simultaneously. Thus, in the United States, the objective of self-sufficiency has been cultivated by the protection of the domestic petroleum industry when the long-term interests of the United States would have best been served by protecting indigenous reserves. At the same time the belief that an ingredient of the industrial economy as ubiquitous as the supply of energy should be cheap has encouraged the sale of fuels at uneconomically low prices; it would have been just as sensible to have argued that if energy is indeed indispensable, it might just as well be expensive.

The case for the realistic pricing of energy, one of the themes of the preceding chapters, is in truth unshakable. To plead for realistic prices is not to advocate free enterprise in preference to some other form of economic organization but merely to ask for a framework within which economic resources can be distributed economically. When the United States administration and the British government (to name only two offenders) arrange that fuels are sold at prices which are uneconomically low, not even the consumers benefit. To be sure, they enjoy the illusion that the natural world is a boundless source of energy. The price they pay is typified by the natural-gas shortages which have afflicted the United States in the past few years and, in Britain, the high cost of providing out of public expenditure or public borrowing the capital needed to finance an industry generating electricity which the over-stimulation of demand has inflated beyond its economic size.

One of the comic aspects of what the governments concerned have been quick to call the energy crisis has been the attempt to foster by exhortation a restraint in the consumption of energy that could more easily be induced by the pursuit of rational pricing policies. Having arranged that water should run uphill, governments have tried unsuccessfully to reverse the process by shouting. What they call the energy crisis will not be over until they recognize, as Canute demonstrated a long time ago, that such exercises are futile. Unhappily, there is a long way to go, at least to judge from the reception accorded by the United States Congress to those parts of President Gerald Ford's proposals, in February 1975, for the abolition of price controls on natural gas and petroleum.

The self-inflicted distortion of the market for energy in many industrialized nations is not, however, a sufficient explanation of the trauma which followed the increase of the price of crude oil during 1973. In no oil-importing industrialized state does the cost of energy exceed 5 per cent of GNP even with oil at 1974 prices. In several, it is a much smaller proportion. On the face of things, industrialized states could expect to pay their bills simply by postponing for a year the benefits of the economic growth which characterized the 1960s. Why then have most of them lapsed into depression?

The first thing to say is that they have been uncommonly shortsighted, and uncommonly ungenerous to each other, in arranging for the equitable distribution of unspent OPEC surpluses among each other. The attempts which have been made so far to create international machinery for recycling the surplus OPEC funds, through the European Commission, the International Monetary Fund and the International Energy Agency of the OECD, are laudable enough. They fail, however, properly to take account of the underlying truth that the economic interdependence which has made possible the rapid growth of the 1960s entails a degree of political interdependence which accords only uneasily with the doctrines of sovereign independence to which all industrial nations still cling. Specifically, countries such as Britain or Italy cannot expect to enjoy the benefits of international trade, and of the recycling of surplus OPEC funds which is necessary if international trade is not to falter, unless they manage their affairs in such a way that their currencies retain their value, which in turn implies that domestic policies are to some extent constrained. Printing money faster than other people is forbidden.

A more serious cause of the troubles of 1974 is that the increase of the price of OPEC oil coincided with the acceleration of inflation in the industrialized West. No doubt, the coincidence is not an accident. Inflation is a measure of the gap between expectations and reality. Although the American discovery that it was possible to appear to pay the bills for an expensive war in South-East Asia by flooding the Eurodollar market with credit was a major cause of the inflation of 1968–74, other industrialized states have been persuaded by the optimism of the 1960s to follow suit. In countries such as Britain, entirely praiseworthy social innovations have for fifteen years been financed partly by the printing presses, and in such a way that industrial investment has been truncated. The belief that energy should as of right be cheap was merely a part of the illusion. When that was dispelled, the weaker brethren inevitably found themselves most seriously handicapped in paying the new bills for oil and in laying out capital to develop new sources of energy.

Pollyannas will say that the increase of the price of oil will turn out to be a blessing in disguise. For it will force on industrial states the policies of fiscal prudence some of them have abandoned in recent years and at the same time compel them to recognize the need for durable financial institutions for the management of international exchanges. That is how the argument goes, but it is only half the truth. Even if the industrialized West as a whole has behaved like a pack of foolish virgins (see Chapter 6), not all states have been equally foolish. In Japan, of all the major industrial states the most dependent on imported oil, GNP fell by 2 per cent between 1973 and 1974 but seems likely to rise by 5 per cent between 1974 and 1975. West Germany appears to be as unperturbed as ever, while France is well on the way to balancing its external books by means of imaginative steps that have been taken to balance the increase cost of imported oil by selling industrial and military equipment to the oil-producing states. Elsewhere, the prospect is less cheerful. In the United States, the administration's budget for 1975–6 was originally built around a deficit of $56 000 million, a sum not very different from the likely unspent surpluses of OPEC states during the same year. In Britain, the government announced in November 1974 that it would finance its operations in the succeeding year by borrowing roughly $20 000 million, mostly from OPEC states. In some countries it is plain the gap between expectation and reality remains to be bridged. In the long run, no doubt, some will swim and some will sink. And while those that sink will no doubt blame the increased price of oil, it is their own devotion to policies of economic self-delusion that will be responsible.

The waywardness of domestic economic policies in many oil-consuming states is only one of the revelations of the period since January 1974. Another is the failure of the oil consumers to act in concert in defence of their common interests, or even to agree where their common interests lie. There are several interrelated issues but the chief of them is the most conspicuous symptom of the energy upheaval, the price of oil. As has been seen, OPEC oil was unreasonably cheap throughout the 1960s, but is now more expensive than the sources of energy that will replace it in the years ahead, in particular nuclear power. If this imbalance persists, both the oil consumers and the oil producers will be damaged, the former because they will be compelled to invest in alternative sources of energy on a larger scale than would otherwise be necessary and the latter because the time will come when there is no ready outlet for their product.

Such declarations are naturally laden with all kinds of political implications, which no doubt explains why most oil consumers have shied away from them (the most conspicuous exception is the United States) and why most oil producers and most developing countries, even those severely handicapped by the present price of oil, regard them as manifestations of commercial colonialism. The waste of economic resources that will be entailed by a prolongation of the regime of high-priced oil initiated by OPEC in 1974 is, however, potentially so great that everybody's interest will be served by an open attack on the anomalies that now persist.

What, then, should be done? In round numbers and in 1974 prices, the equitable price of OPEC oil should be about $7.00 a barrel, compared with something in excess of $10.00 a barrel fixed by the uniform pricing arrangements introduced at the end of 1974. This is what OPEC oil will cost ten years or so from now, when the oil consumers have invested in alternative sources of energy. But if the oil producers wait until their customers have laid out capital on such a scale that imports are no longer necessary, they will find that the new sources of energy are commercially protected and that their own revenues are much reduced. Some OPEC members, Kuwait for example, would not be seriously affected. Others, Iran and Algeria, would be in serious difficulties, given their likely need of funds for economic and social development. In the circumstances, it is neither neo-colonialist nor merely selfish that the oil consumers should take steps to bring down the price of oil. The policies that should be followed are also clear. First, even as a temporary expedient, the oil consumers must reduce their oil consumption by between 10 and 15 per cent – enough to bring the divergent interests of OPEC members to the surface. Second, they must hold clearly in mind, and make public whenever necessary, an estimate of what they consider to be an equitable price for oil. Seven dollars a barrel is as good as anything. If they can find alternative sources of energy more cheaply, they will invest accordingly and OPEC sales will be correspondingly reduced. If they are forced to pay more for energy over any substantial length of time, they will protect their investments with import duties, quotas and the whole apparatus of commercial regulation, and OPEC's sales of oil will be more drastically affected. As it happens, this is precisely the policy advocated by Dr Henry Kissinger, the US Secretary of State in the closing weeks of 1974. To begin with, at least, he is a prophet in the wilderness.

The industrialized nations of the West are in reality a motley crew. Some of them, Canada for example, are also oil exporters, and calculate that in the short run at least they have more to gain from the export taxes now levied on oil and natural gas than they have to lose from the uneconomic pattern of investment likely to be engendered if the high price of oil persists. The trouble, of course, is that such policies lay them open to precisely the risks that OPEC states now run. Other oil consumers, Britain for example, consider themselves to be so near the point of self-sufficiency in oil that they have no stomach for a strictly commercial bargain with OPEC; where they miscalculate is in their supposition that it will be more profitable to drill holes in the more distant parts of the North Sea to produce ten-dollar oil than to pay cash for cheaper OPEC oil in the expectation that some at least of what they spend will be returned by the flow of international trade.

Fortunately, these attitudes will not persist. By the end of 1974, there were signs that the narrow and contradictory self-interests of the different oil consumers were being submerged in economic logic. For the industrialized West, the high price of oil is no great hardship, at least if the economies of the West are managed well. What does matter is that unreasonably high prices will force on the West patterns of industrial investment which are premature and a wasteful diversion of economic resources. This is why the objectives of the oil consumers should be to reduce the price of oil to something like \$7.00 a barrel.

The steps which oil-consuming states must take to restore this happy state of affairs are plain. The rapid development of alternative sources of energy, even in modest amounts, is one obvious step. Saving energy is another. But because substantial progress in these directions is certain to be slow, since capital investment in such enterprises is necessarily slow, there is also a strong case why the oil consumers should impose on themselves measures to conserve energy that go beyond economic commonsense (see Chapter 8), if only for a time. A few years of petrol rationing would be a small price to pay for a speedy return to a rational international market for petroleum. The practical difficulty is that such a strategy could succeed only if the oil consumers were to act in concert. This, no doubt, is why the United States Administration had begun, towards the end of 1974, to press for a coordinated policy on oil consumption with France and Norway standing aloof. The notion that OPEC members should at the same time be offered a permannently guaranteed price for their exports of crude oil, put forward by Dr Henry Kissinger, is an entirely sensible accompaniment of this policy. The oil importers have nothing to lose by undertaking that in the long run the international price of oil will be substantially the same as that of energy from the alternative sources now to hand, while OPEC members may derive some benefit, more psychological than economic, therefrom.

The reasons why the good sense of this policy was not immediately recognized by the oil consumers, in the months of anxious discussion early in 1974, are still obscure. Some have responded to the transformation of the petroleum market like rabbits in the headlights of a car. Others have hidden from the truth behind illusions of different kinds – the Nixon Administration's policy of self-sufficiency in energy, the British government's hopes of cheap oil from the North Sea and the French government's belief that it could turn the course of events to its own advantage by increasing its exports to OPEC states. But the most serious obstacle to progress has been the unwillingness of the oil consumers to recognize openly among themselves what the objectives of their policy should be. The

OPEC oil embargoes of 1973 have made their mark. The governments of the oil consumers repeatedly declare they have no wish for confrontation with OPEC as a whole, which is sensible enough, and then go on to suppose that that prescribes their right to strike an honest bargain.

In the long run, neither OPEC members nor the oil consumers will benefit from such reticence. The interests of the oil-producing states, which have since the begining of 1974 enjoyed unprecedented incomes from the export of oil, will in the long run be best served if there is a substantial market for what they have to sell, at a reasonable price. If, in the two or three years during which the price of oil is likely to remain above the sensible economic level, they succeed in accumulating substantial credits with the oil consumers (the Morgan Guarantee Trust estimated at the end of 1974 that the accumulated surpluses of OPEC states by the end of the 1970s would amount to a little less than $300 000 million) so much the better. A group of states for which the prospects of rapid economic and social progress have not in the past been bright will have joined the ranks of the well-to-do. It is neither paradoxical nor ironic that the oil-exporting states will find it necessary, in the years ahead, to join the industrialized states of the West in planning to augment their own supplies of energy by building nuclear power stations.

So it is that decades from now the energy crisis of 1973–4 may turn out to seem not a cataclysm for the West but a period during which the benefits of the continuing industrial revolution came to be more widely shared. That it will also seem the point at which oil consumers first clearly recognized that their consumption of petroleum must in due course be attenuated is less remarkable. That lesson could have been learned at any time in the past two decades. And if it should turn out, in the years ahead, that the recent argument about the price of oil is mirrored by arguments about the price of uranium, or some other important raw material, there is at least a chance that recent events will have shown us all that these are merely occasions for the exercise of that blend of courage and realism on which industrial society is founded.

Bibliographical notes

The following is a brief guide to some particularly helpful books and documents which may be consulted for further information. They are arranged roughly in the order in which the topics with which they deal occur in the text.

Colin Clark, *Population Growth and Land Use*, Macmillan (London), 1967 and Colin Clarke and M. R. Haswell, *The Economics of Subsistence Agriculture*, Macmillan (London) 1964. Between them, these books, although largely concerned with the food supplies of prehistoric and better known societies, contain some of the most explicit calculations of the energy requirements and the economics of ancient agriculture.

David S. Landes, *The Unbound Prometheus*, Cambridge University Press, 1969. Among the vast literature of the Industrial Revolution, this book stands out for its attempt to tell the economic history of technological innovation.

W. S. Jevons, *The Question of Coal*, Macmillan (London), 1861. Jevons's fears for the future of British coal are lucidly expressed, so that his book (which provoked the establishment of a Royal Commission) is still among the more cogent examples of doomsday literature.

Energy in the World Economy, Joel Darmstadter, Johns Hopkins Press for Resources for the Future Inc., 1970. Although this compilation of statistics stops short at 1968, it remains the most convenient way of tracing the relationship between economic growth and energy consumption in different countries. One of its defects is that it does not distinguish between different forms of energy consumption except in the simplest way.

Oil: The Present Situation and Future Prospects, OECD,

1973. Among the attempts of international organizations to forecast the future demand for energy and in particular oil, this is distinguished by the importance attached to price.

Resources for America's Future, Johns Hopkins Press for Resources for the Future Inc., 1965. This monumental study of the course of demand for raw materials, including energy, in the United States has the virtue that the exercise is based not on the extrapolation of past trends but on an attempt to estimate what would happen in different sections of the market.

The 1970 National Power Survey (5 parts), Federal Power Commission, Washington, 1972. A detailed anatomy of the electrical power industry in the United States.

David Pimentel *et al.*, 'Food Production and the Energy Crisis', *Science, 182*, 443, 1973. An attempt to construct an energy balance sheet for intensive agriculture in the United States which has been variously interpreted as a proof that modern agriculture is, in one sense or another, too energy-intensive and as a proof that it is still worthwhile using large quantities of fertilizer to grow corn.

Resources and Man, National Academy of Sciences, W. H. Freeman. This document contains a succinct version of Dr R. King Hubbert's now classic argument of the likely pattern of exhaustion of petroleum reserves. Even those who consider his numerical estimates to be too gloomy accept the principles.

US Mineral Resources, Geological Survey Professional Paper 820, US Government Printing Office, 1973. One of the most optimistic of recent estimates of energy reserves extending outside the United States, which has been criticized since publication by the US Environmental Protection Agency and by a committee of the US National Academy of Sciences with Dr R. King Hubbert as chairman.

The Increased Cost of Energy – Implications for UK Industry, NEDO, 1974. An estimate of the economic impact of higher oil prices on different sectors of British industry.

Finance for the Nationalised Industries, UK White Paper, HMSO 1967. The document which spelled out the rules by which nationalized industries in Britain should determine prices, a policy abandoned within three years.

M. A. Adelman, *The World Petroleum Market*, Johns Hopkins Press for Resources for the Future Inc., 1972. A controversial book whose chief argument is that the price of oil in the 1950s and 1960s was too high, but which is one of the best sources of information on costs of oil production in different parts of the world.

US Energy Outlook, National Petroleum Council, 1972. The outcome of a computer forecast of energy supply and demand which makes explicit the costs of alternative sources of energy as well as calculations of the minimum prices that would justify these investments.

E. N. Tiratsoo, *Oilfields of the World*, Scientific Press, 1973. A thorough survey of the physical characteristics of individual oil fields lacking in economic information.

Final Environmental Statement for the Prototype Oil Shale Leasing Programme (6 vols), US Dept of the Interior, 1973. A handy compilation of geological, technical and economic information about the oil shale deposits of the United States as well as an account of the environmental problems exploitation will cause.

Michael V. Posner, Fuel Policy: *A Study in Applied Economics*, Macmillan (London), 1973. A dissection of British energy policy in the past two decades which takes account of the social cost of energy exploitation.

The Potential for Energy Conservation, Executive Office of the President, US Government Printing Office, 1973. Contains invaluable detailed information on the physical and economic possibilities for energy economy.

First Interim Report of Ford Foundation Energy Commission, August 1974. A mouse of a document, arguing the case for Zero Energy Growth but for practical purposes demolished by the hostile criticisms of members of the project's Advisory Board printed as an Appendix.

Cornell Workshops, Report printed in evidence to the US Joint Congressional Committee on Atomic Energy on Energy Research and Development, March 1974. A careful analysis of the research and development needed in the United States to regain energy self-sufficiency.

Index

Compiled by Gordon Robinson